Springer Theses

Recognizing Outstanding Ph.D. Research

Aims and Scope

The series "Springer Theses" brings together a selection of the very best Ph.D. theses from around the world and across the physical sciences. Nominated and endorsed by two recognized specialists, each published volume has been selected for its scientific excellence and the high impact of its contents for the pertinent field of research. For greater accessibility to non-specialists, the published versions include an extended introduction, as well as a foreword by the student's supervisor explaining the special relevance of the work for the field. As a whole, the series will provide a valuable resource both for newcomers to the research fields described, and for other scientists seeking detailed background information on special questions. Finally, it provides an accredited documentation of the valuable contributions made by today's younger generation of scientists.

Theses may be nominated for publication in this series by heads of department at internationally leading universities or institutes and should fulfill all of the following criteria

- They must be written in good English.
- The topic should fall within the confines of Chemistry, Physics, Earth Sciences, Engineering and related interdisciplinary fields such as Materials, Nanoscience, Chemical Engineering, Complex Systems and Biophysics.
- The work reported in the thesis must represent a significant scientific advance.
- If the thesis includes previously published material, permission to reproduce this must be gained from the respective copyright holder (a maximum 30% of the thesis should be a verbatim reproduction from the author's previous publications).
- They must have been examined and passed during the 12 months prior to nomination.
- Each thesis should include a foreword by the supervisor outlining the significance of its content.
- The theses should have a clearly defined structure including an introduction accessible to new PhD students and scientists not expert in the relevant field.

Indexed by zbMATH.

More information about this series at http://www.springer.com/series/8790

Kenichi Endo

Kinetically Controlled Stepwise Syntheses of a Heterometallic Complex and a Tetrahedral Chiral-at-Metal Complex

Doctoral Thesis accepted by the University of Tokyo, Tokyo, Japan

Springer

Author
Dr. Kenichi Endo
Institute for Chemical Research
Kyoto University
Gokasho, Kyoto, Japan

Supervisor
Prof. Mitsuhiko Shionoya
Department of Chemistry, Graduate School
of Science
The University of Tokyo
Bunkyo-ku, Tokyo, Japan

ISSN 2190-5053 ISSN 2190-5061 (electronic)
Springer Theses
ISBN 978-981-16-1162-9 ISBN 978-981-16-1163-6 (eBook)
https://doi.org/10.1007/978-981-16-1163-6

This Springer imprint is published by the registered company Springer Nature Singapore Pte Ltd.
The registered company address is: 152 Beach Road, #21-01/04 Gateway East, Singapore 189721, Singapore

Supervisor's Foreword

All elements in the periodic table can be stereo-centered through chemical bonds. Like carbon atoms, metal elements that occupy about 80% of the periodic table can be structurally chiral centers. That is, it is possible to design and synthesize a chiral-at-metal complex having chirality at the metal center using only an achiral ligand. A coordinate bond is a directional, reversible bond formed by a metal and a donor atom, and its thermodynamic and kinetic stability depends on the type of metal, the oxidation number, and the type of donor atom. Metals can be the key to exhibiting structural and spatially specific reactions and anisotropic physical properties in chiral-at-metal complexes and their aggregates with various coordination structures.

Chiral-at-metal complexes with metal-centered chirality and their self-assembled metal complexes can provide a field for asymmetric reactions, chiral spaces, and anisotropic physical properties that are specific to unique molecular structures and assembly modes. The development of "chemistry of asymmetric metals" requires an asymmetric induction principle different from that of asymmetric carbons, which can control the dynamic properties of coordination bonds that govern the substitution activity and stability of metal complexes. The purposes of our study on chiral-at-metal complexes are to develop an asymmetric induction method for metal centers based on control of ligand substitution activity by molecular design and setting of chemical environment, and to search for methods of constructing chiral metal catalysts, chiral functional spaces, and anisotropic substances using of molecules with asymmetric metals.

Our research group started various studies in 2016 focusing on "coordination asymmetry." As part of this, Dr. Kenichi Endo challenged the synthesis of a Werner-type tetrahedral chiral-at-zinc complex with a stable molecular configuration in his doctoral course and succeeded in applying it to complete asymmetric induction and asymmetric catalytic reactions. This study greatly expands the possibilities of tetrahedral chiral-at-metal complexes, and the achievement was published in Nature Communications in 2020. He was also awarded the Best Student Presentation Award in the 69th Conference of Japan Society of Coordination Chemistry and the School of

Science Encouragement Award from Graduate School of Science, the University of Tokyo. I hope that this book will provide a rational design principle for chiral-at-metal catalysts that expands the scope of "chemistry of asymmetric metals."

Tokyo, Japan
January 2021

Prof. Mitsuhiko Shionoya

Parts of this thesis have been published in the following journal articles:

- Ube, H.; Endo, K.; Sato, H.; Shionoya, M. Synthesis of Hetero-Multinuclear Metal Complexes by Site-Selective Redox Switching and Transmetalation on a Homo-Multinuclear Complex. *J. Am. Chem. Soc. 2019*, **141**, 10384–10389.
- Endo, K.; Liu, Y.; Ube, H.; Nagata, K.; Shionoya, M. Asymmetric Construction of Tetrahedral Chiral Zinc with High Configurational Stability and Catalytic Activity. *Nat. Commun. 2020*, **11**, 6263.

Acknowledgments

This research was conducted under the supervision of Prof. Dr. Mitsuhiko Shionoya (the University of Tokyo) and Assistant Professor Dr. Hitoshi Ube (the University of Tokyo). The single-crystal XRD analysis in Chap. 2 was conducted in collaboration with Dr. Hiroyasu Sato (Rigaku Corporation). The experiments in Chap. 3 were conducted in collaboration with Mr. Yuanfei Liu (master student, currently PhD student) and Project Assistant Professor Dr. Koichi Nagata (the University of Tokyo, currently Tohoku University). I sincerely appreciate their assistance on this research.

Professor Dr. Mitsuhiko Shionoya enthusiastically motivated me with scientific curiosity throughout my Ph.D. course and gave me many fruitful suggestions and corrections on the research. The research target in Chap. 3 originated from his idea of "tetrahedral chiral-at-metal complex." I would like to express my profound gratitude here.

Assistant Professor Dr. Hitoshi Ube advised me so many times in a daily level both on research and on life in this group. The unsymmetrical ligand **L1** used in Chap. 2 was a derivative of a prototype ligand born in a discussion with him. I am deeply grateful to him.

Project Assistant Professor Dr. Koichi Nagata taught me numerous ideas and experimental techniques, especially air-free techniques necessary for Chap. 3. Meanwhile, his devotion to research motivated me a lot. I am deeply grateful to him.

Mr. Yuanfei Liu helped me on the research topic in Chap. 3. His diligence strongly promoted this research. I also learned as many things by teaching him as he learned. I am very thankful to him.

Associate Professor Dr. Shohei Tashiro and Assistant Professor Dr. Yusuke Takezawa suggested me useful advice and maintained the laboratory apparatus. I appreciate their cooperation.

The colleagues in Shionoya Group helped me with daily troubles and cooperatively built up the joyful research environments. I was happy to work as a colleague. Thanks everyone!

Professor Dr. Eric Meggers and the colleagues in Philipps-Universität Marburg, Germany, kindly accepted my short research internship and gave me a unique opportunity to touch with their research, which was closely related to the topic in Chap. 3. I appreciate them very much.

Advanced Leading Graduate Course for Photon Science (ALPS) Program, JSPS Research Fellowship for Young Scientists DC1, and Graduate Research Abroad in Science Program (GRASP) offered me financial supports. I appreciate these programs.

Finally, I would like to show my gratitude to my wife, who supported my daily life and also sometimes even this research.

Contents

Abbreviations

a, b, c, α, β, γ	Lattice constants
Abs.	Absorbance
Ac	Acetyl
ATR	Attenuated total reflection
A.U.	Arbitrary unit
bpy	2,2′-bipyridine
br	Broad
Bu	Butyl
calcd.	Calculated
CCDC	Cambridge Crystallographic Data Centre
CD	Circular dichroism
COSY	Correlation spectroscopy
d	Doublet
δ	Chemical shift
D_{calc}	Calculated density
de	Diastereomeric excess
dr	Diastereomer ratio
d-Fourier	Differential Fourier
DMF	*N*,*N*-dimethylformamide
DMSO	Dimethyl sulfoxide
ee	Enantiomeric excess
eq.	Equivalent
equiv.	Equivalent
ESI	Electrospray ionization
Et	Ethyl
FT	Fourier transform
Fw	Formula weight
g_{em}	Dissymmetry factor for emission
GOF	Goodness of fit
HMDSO	Hexamethyldisiloxane
HPLC	High-performance liquid chromatography
HR	High resolution

^{i}	Iso-
IR	Infrared
J	Coupling constant
K_f	Formation constant
l	Cell length
L	Ligand
λ	Wavelength
m	Multiplet
m	Mass
M	Metal
M	Formula weight
μ	Linear absorption coefficient
Me	Methyl
Mes	Mesityl
m.p.	Melting point
MS	Mass spectrometry
MS	Molecular sieves
^{n}	Normal-
n	Normal
NBS	*N*-bromosuccinimide
NIR	Near infrared
NMR	Nuclear magnetic resonance
NOE	Nuclear Overhauser effect
NOESY	Nuclear Overhauser effect correlated spectroscopy
ORTEP	Oak ridge thermal ellipsoid plot
p	*Para*
p	Decimal cologarithm of
Ph	Phenyl
ppm	Parts per million
Pr	Propyl
θ	Angle
Θ	Weiss constant
ρ_{calc}	Calculated density
R_1	Reliability factor 1
R_{int}	Internal consistency of equivalents
RT	Room temperature
s	Singlet
SC-XRD	Single-crystal X-ray diffraction
t	Triplet
^{t}	*Tert-*
T	Temperature
T_C	Curie temperature
Tf	Trifluoromethanesulfonyl
TFA	Trifluoroacetic acid
THF	Tetrahydrofuran

TMS	Tetramethylsilane
TOF	Time-of-flight
Ts	*p*-toluenesulfonyl
UV	Ultraviolet
V	Volume
vis	Visible
wR_2	Weighted reliability factor 2
XPS	X-ray photoelectron spectroscopy
XRD	X-ray diffraction
z	Charge number
Z	Number of formula units in the unit cell

Chapter 1
General Introduction

Abstract Metal complexes form a class of compounds which are attractive for their rich physicochemical properties. Syntheses of metal complexes are usually based on direct reaction of metal sources and ligands. However, this one-step method is not always useful because multiple products are sometimes generated. One method to overcome this problem is to elaborate the system so that one product is thermodynamically much favored than other products, but such thermodynamically stable product is limited in variation. In contrast, stepwise synthesis can provide a variety of structures with high selectivity through kinetically controlled pathways. This method has been widely employed for some inert metal complexes, but rarely used for labile metal complexes. Therefore, this thesis aimed at development of kinetically controlled stepwise synthesis of complexes of labile metal ions, which would contribute to a greater freedom of the design of metal complexes.

Keywords Metal complexes · Selective synthesis · Kinetic control

1.1 Selective Synthesis of Metal Complexes

Metal complexes, comprised of metal atom(s) and coordinating ligand(s), reside in the cross section of organic and inorganic chemistry fields. Metal complexes exhibit many attractive chemical and physical properties based on the combination of the diverse properties of metal elements with the high designability of organic ligands. For example, metal complexes show unique assembled structures (host compounds, porous materials), physical properties (optical properties, magnetism, conductivity), and chemical reactivities (catalysis, photosensitization, enzyme mimicry) [1, 2].

To synthesize a metal complex, typically, a direct one-step reaction between the inorganic part (metal source) and the organic part (ligands) of the target complex is employed (Fig. 1.1a). This process, called complexation, is usually rapid and under thermodynamic control because metal–ligand coordination bonds are labile. However, in some cases with heteroleptic, heterometallic, or isomeric complexes, the products can be a cumbersome mixture of multiple complexes (Fig. 1.1b). In such a

K. Endo, *Kinetically Controlled Stepwise Syntheses of a Heterometallic Complex and a Tetrahedral Chiral-at-Metal Complex*, Springer Theses,
https://doi.org/10.1007/978-981-16-1163-6_1

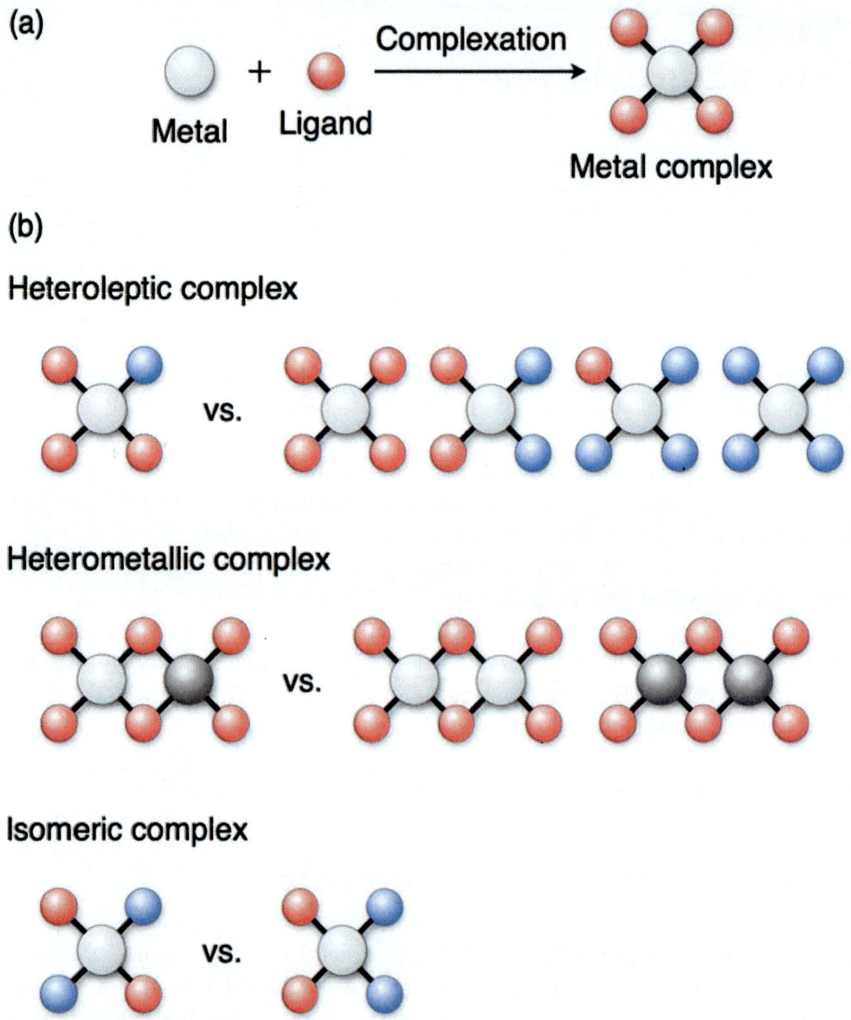

Fig. 1.1 **a** Schematic diagram of typical metal complexation. **b** Examples of metal complexes whose synthesis can produce multiple products

case, development of a selective synthesis method is desired to spare the difficulty and labor of separation.

One method to achieve selective synthesis is to modify the system so that the desired product will be thermodynamically more stable than the other possible products. Then, the desired complex can be selectively obtained by thermodynamically controlled one-step synthesis (Fig. 1.2a). This can be realized by elaboration of organic ligands [3, 4], metals with complementary properties [5], and/or specific interactions with counterions [6, 7] or guest species [8] (Fig. 1.2b). For example, selective syntheses of some heteroleptic [3, 4], heterometallic [5], and isomeric [6, 7]

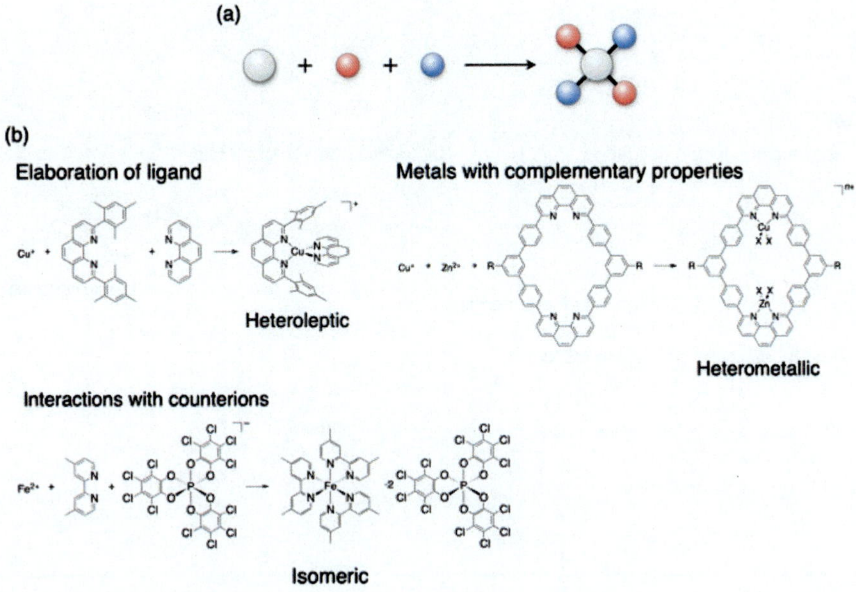

Fig. 1.2 Thermodynamically controlled one-step synthesis of metal complexes. **a** Schematic representation. **b** Examples where selective syntheses of heteroleptic, heterometallic, and isomeric complexes were conducted based on elaboration of ligand [4], metals with complementary properties (X = a solvent molecule or counterion) [5], and interactions with counterions [7], respectively

complexes based on this method are reported (Fig. 1.2b). However, the requirement of thermodynamic stability poses limitations on the design of ligands and the choice of metals.

An alternative method is to divide the synthesis process into a few steps to enable kinetic control, just as organic synthesis does (Fig. 1.3a) [9]. This method includes sequential addition of different ligands [10] or metals [11], post-modification of metal redox states or ligand structures after complexation [12], and temporary introduction of auxiliary component followed by its removal [13, 14] (Fig. 1.3b). In every case, the initial steps are under thermodynamic control, but the final step is under kinetic control to give a different selectivity from thermodynamic control. Here, a selective reaction under kinetic control is easier than the one-step complexation because the final step will involve less changes in the structure of the complex and require milder conditions. For example, selective syntheses of the heteroleptic [10], heterometallic [11, 12], and isomeric [13, 14] complexes which cannot be selectively obtained by the thermodynamic one-step method have been reported (Fig. 1.3b).

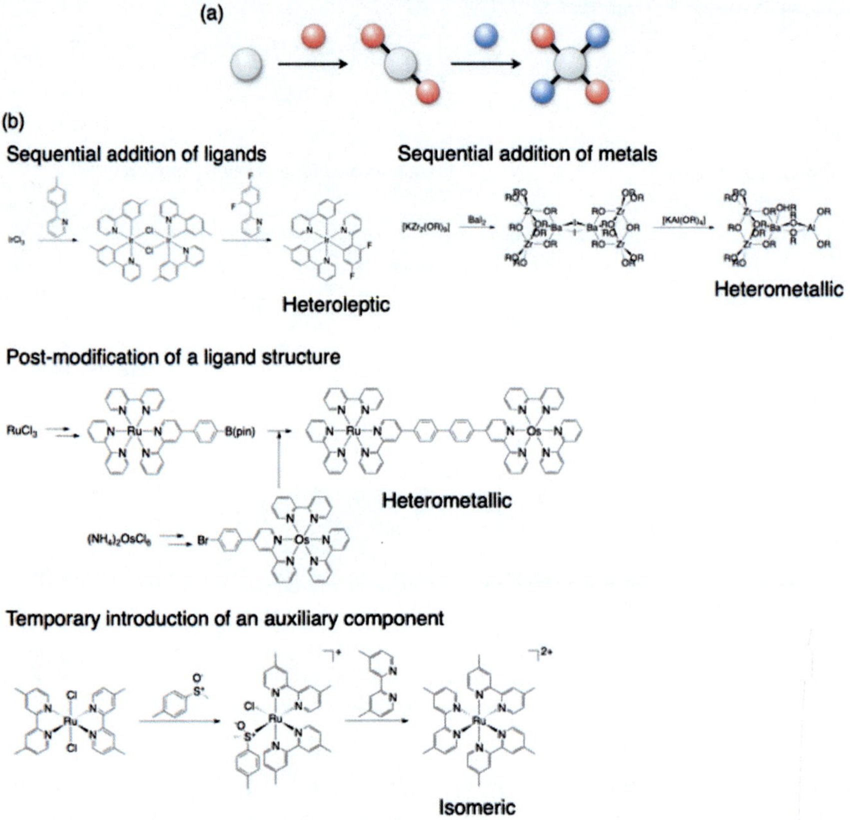

Fig. 1.3 Kinetically controlled stepwise syntheses of metal complexes. **a** Schematic representation. **b** Examples based on sequential addition of ligands [10] or metals [11], post-modification of a ligand structure [12], and temporary introduction of an auxiliary component [14]

1.2 Target Scope of Kinetically Controlled Stepwise Synthesis

Despite its utility, the target scope of kinetically controlled stepwise synthesis has been severely limited. Previous examples are focused mainly on 4d or 5d metals (Fig. 1.3b) [10–12, 14], octahedral Cr^{III} [15] or Co^{III} [16], or non-Werner-type complexes [17] (Fig. 1.4). This is because these complexes are relatively inert [18], which makes kinetic control easier.

Other kinds of metal complexes are usually more labile [18], so kinetic control is more difficult. Only a limited number of examples have been reported: lanthanide complexes with octadentate ligands [9, 19], Cu^{I} complexes with bulky ligands [20], and polynuclear complexes [21–23] (Fig. 1.5).

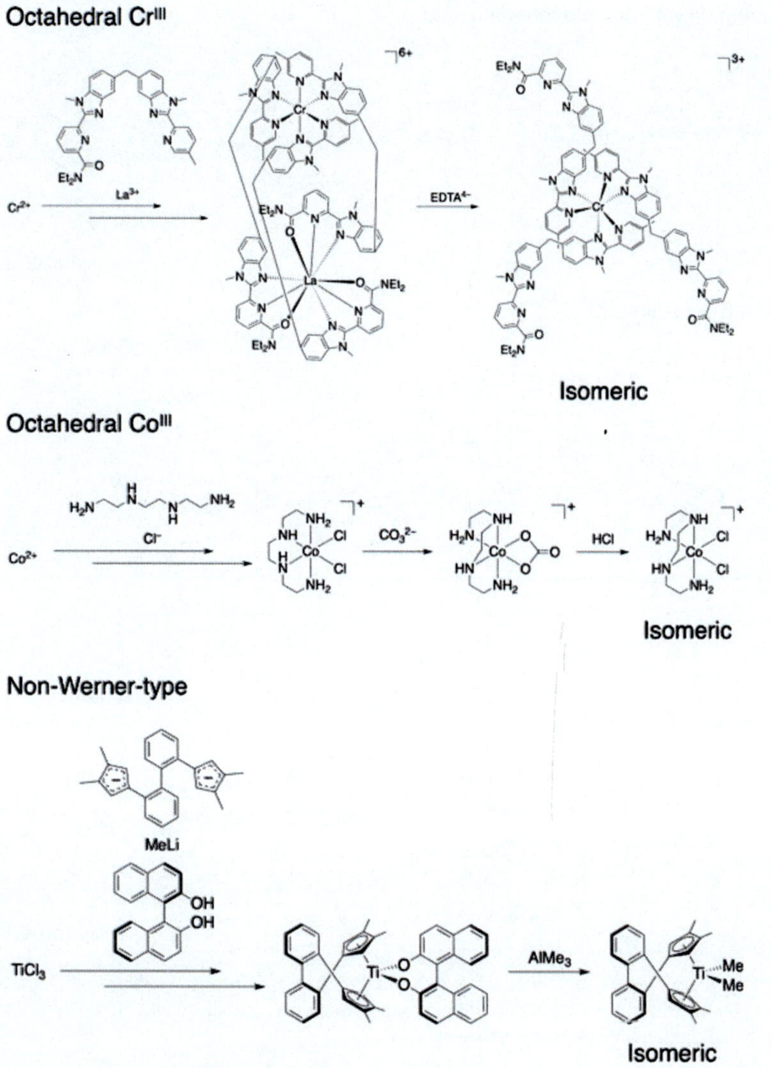

Fig. 1.4 Examples of the kinetically controlled stepwise selective syntheses of octahedral Cr^{III} [15] and Co^{III} [16] and non-Werner-type [17] complexes

Therefore, it is desired to expand the target scope of kinetically controlled stepwise synthesis for complexes of labile metal ions, by developing new methodologies and design principles. Such development will provide a greater freedom to the design of metal complexes, which can be used as functional chemicals.

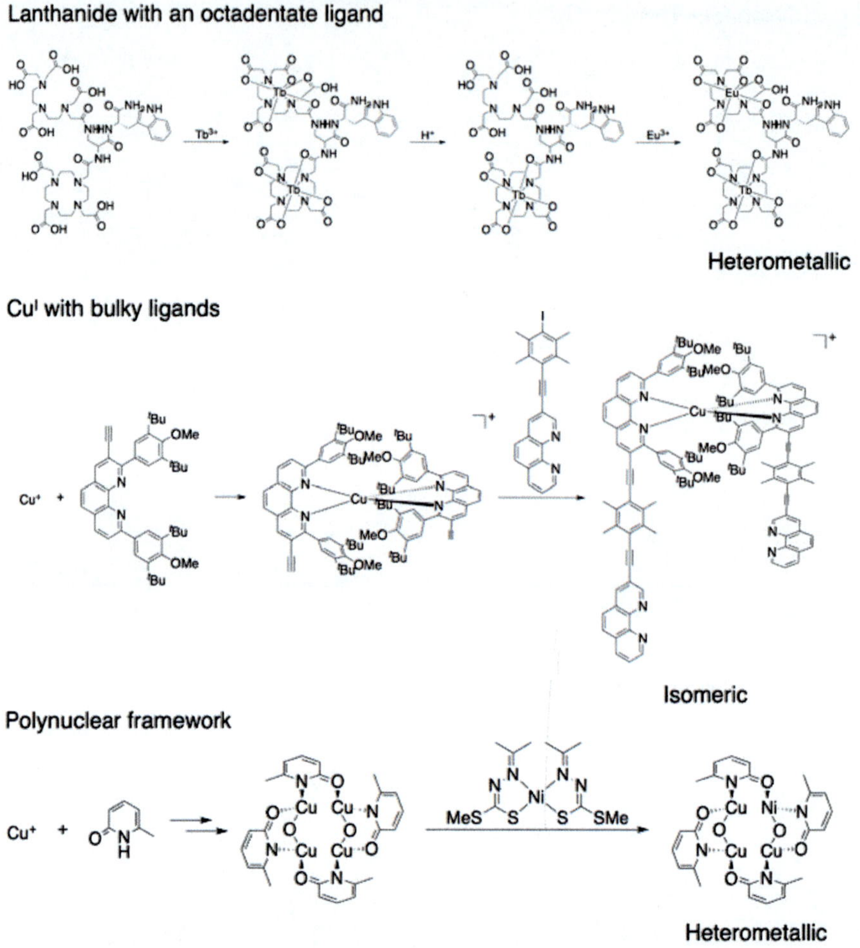

Fig. 1.5 Examples of the kinetically controlled stepwise selective syntheses of Werner-type complexes of other metals based on lanthanide with an octadentate ligand [19], Cu^{I} with bulky ligands [20], and a polynuclear framework [23]

1.3 Overview of This Study

In this work, I aimed at development of kinetically controlled stepwise synthesis of complexes of labile metal ions in two distinct fields (Fig. 1.6): (i) a heterometallic complex and (ii) a chiral-at-metal complex, which is a particular kind of an isomeric complex.

For the heterometallic complex, I propose "site-selective redox switching and transmetalation" as a new stepwise method and applied it to a labile Co^{II} and Ni^{II} complex. As a result, selective synthesis of the heterometallic complex $[Co^{II}Ni^{II}_{3}\mathbf{L1}_{3}X_{6}]$ (X = solvent molecule or counteranion) was achieved in four

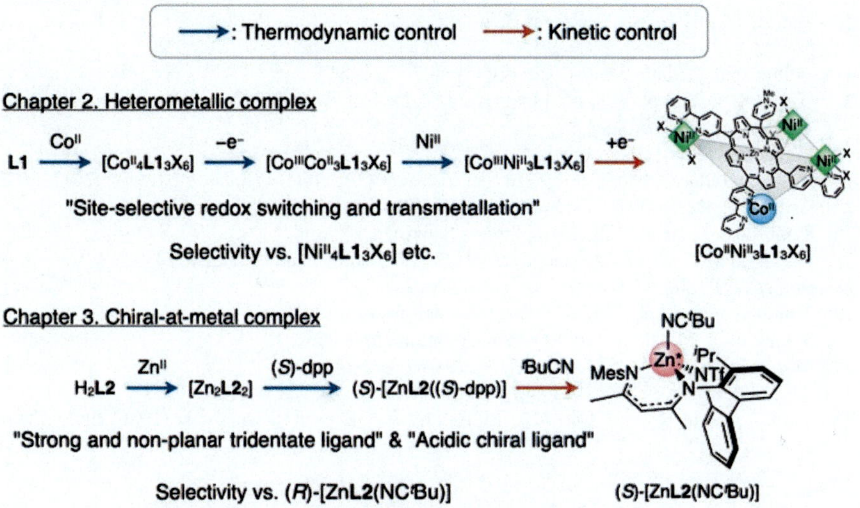

Fig. 1.6 Summary of this study. For abbreviations, refer to each chapter

steps. This method showed kinetically controlled selectivity against a thermodynamic mixture of complexes such as $[Ni^{II}_4\mathbf{L1}_3X_6]$. This topic is discussed in Chap. 2.

For the chiral-at-metal complex, I propose the use of a strong and non-planar tridentate ligand in combination with acidic chiral ligand and applied it to a tetrahedral Zn^{II} complex, which is typically highly labile. As a result, selective synthesis of the chiral-at-metal complex (*S*)-[Zn**L2**(NC^{t}Bu)] was achieved in three steps. This strategy gave kinetically controlled selectivity against a thermodynamic mixture with (*R*)-[Zn**L2**(NC^{t}Bu)]. This topic is discussed in Chap. 3.

These ideas would be useful in preparation of new heterometallic or chiral-at-metal complexes which have been difficult to achieve. Such complexes can be expected for unique physical or chemical properties. As an example, enantioselective catalysis by the resultant complex is demonstrated in the latter topic.

References

1. Gispert JR (2008) Coordination chemistry. WILEY-VCH, Weinheim
2. Hasegawa Y, Ito H (2014) 錯体化学—基礎から応用まで. Kodansha, Tokyo
3. De S, Mahata K, Schmittel M (2010) Chem Soc Rev 39:1555–1575
4. Schmittel M, Ganz A (1997) Chem Commun 999–1000
5. Kuritani M, Tashiro S, Shionoya M (2012) Inorg Chem 51:1508–1515
6. Lacour J, Moraleda D (2009) Chem Commun 7073–7089
7. Lacour J, Jodry JJ, Ginglinger C, Torche-Haldimann S (1998) Angew Chem Int Ed 37:2379–2380
8. Johnson DW, Raymond KN (2001) Supramol Chem 13:639–659
9. Sorensen TJ, Faulkner S (2018) Acc Chem Res 51:2493–2501

10. Tamura Y, Hisamatsu Y, Kumar S, Itoh T, Sato K, Kuroda R, Aoki S (2017) Inorg Chem 56:812–833
11. Heidemann T, Mathur S (2017) Inorg Chem 56:234–240
12. Welter S, Salluce N, Belser P, Groeneveld M, De Cola L (2005) Coord Chem Rev 249:1360–1371
13. Meggers E (2010) Chem Eur J 16:752–758
14. Hesek D, Inoue Y, Ishida H, Everitt SRL, Drew MGB (2000) Tetrahedron Lett 41:2617–2620
15. Cantuel M, Bernardinelli G, Muller G, Riehl JP, Piguet C (2004) Inorg Chem 43:1840–1849
16. Sargeson AM, Searle GH (1967) Inorg Chem 6:787–796
17. Ringwald M, Stürmer R, Brintzinger HH (1999) J Am Chem Soc 121:1524–1527
18. Richens DT (2005) Chem Rev 105:1961–2002
19. Tremblay MS, Sames D (2006) Chem Commun 4116–4118
20. Schmittel M, Mahata K (2008) Chem Commun 2550–2552
21. Carnes ME, Collins MS, Johnson DW (2014) Chem Soc Rev 43:1825–1834
22. Davies G, El-Sayed MA, El-Toukhy A (1992) Chem Soc Rev 21:101–104
23. Cai GZ, Davies G, El-Sayed MA, El-Toukhy A, Gilbert TR, Onan KD (1986) Inorg Chem 25:1935–1940

Chapter 2
Heterometallic $Co^{II}Ni^{II}_3$ Complex

Check for updates

Abstract Heterometallic multinuclear complexes have great potential to exhibit unique chemical and physical properties based on the cooperation of different metal ions. However, most examples previously reported depend on one-step methods or are based only on 4d- or 5d-metal ions. Here, a new strategy toward an otherwise difficult arrangement of different 3d-metal ions has been achieved by kinetically controlled multistep sequence: site-selective redox switching and transmetalation. This strategy utilizes inequivalent metal sites and a redox-active metal ion. As a model study, I chose a trigonal-pyramid-shaped tetranuclear structure to generate inequivalent metal sites, cobalt(II) as a redox-active metal ion, and nickel(II) as a second metal ion, which is difficult to distinguish with cobalt by thermodynamic methods. The construction of heterometallic $Co^{II}Ni^{II}_3$ complex was accomplished in four-step sequence: (1) complexation with Co^{II}; (2) site-selective oxidation at one of the Co^{II} sites; (3) site-selective transmetalation of the other Co^{II} sites with Ni^{II}; (4) reduction of the Co^{III} site. Notably, the obtained heterometallic complex could not be synthesized directly from ligand, Co^{II}, and Ni^{II} under thermodynamic control.

Keywords Heterometallic complex · Polynuclear complex · Self-assembly · Site-selective oxidation · Transmetalation

2.1 Introduction

Heterometallic complexes are complexes containing several different metal elements. Such complexes are attractive because interactions between the metal atoms can lead to unique chemical and physical properties such as luminescence [1–4], magnetism [3, 5, 6], and catalysis [3, 7, 8] (Fig. 2.1a–c). Heterometallic complexes are also seen in biological systems, especially in metalloenzymes (Fig. 2.1d) [9, 10].

To obtain a desired heterometallic complex, complexation between ligands and several different metal sources should be controlled [11, 12]. If no care was taken, a complex with an undesired number of metal atoms or an isomer with an undesired positioning of the metal atoms can be generated (Fig. 2.2a). Conventionally, thermodynamic control has been widely employed to achieve selective formation

K. Endo, *Kinetically Controlled Stepwise Syntheses of a Heterometallic Complex and a Tetrahedral Chiral-at-Metal Complex*, Springer Theses,
https://doi.org/10.1007/978-981-16-1163-6_2

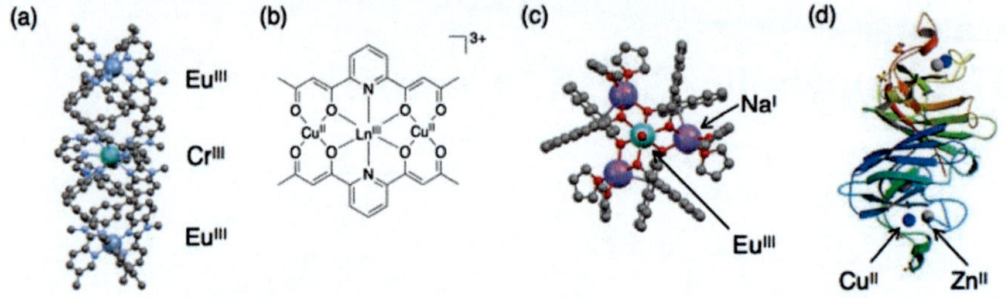

Fig. 2.1 Examples of heterometallic complexes. **a–c** Heterometallic complexes which exhibit **a** luminescence [4], **b** magnetism [6], and **c** catalytic activity [8]. **d** Heterometallic enzyme, Cu-Zn human superoxide dismutase (PDB: 2C9V) [10]

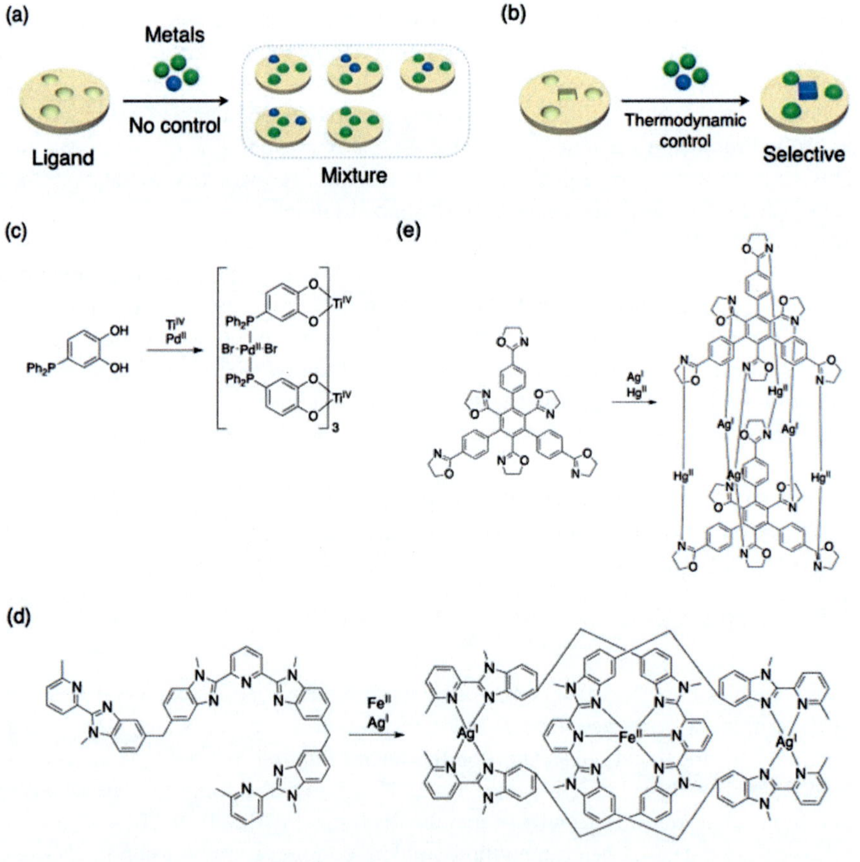

Fig. 2.2 Synthesis of heterometallic complexes. **a, b** Typical schemes of **a** non-selective synthesis without any control and **b** selective synthesis under thermodynamic control. **c–e** Examples of selective syntheses under thermodynamic control based on **c** hard/soft ligands and metals [13], **d** different preferred coordination numbers/geometry [14], and **e** different ionic charges [15]

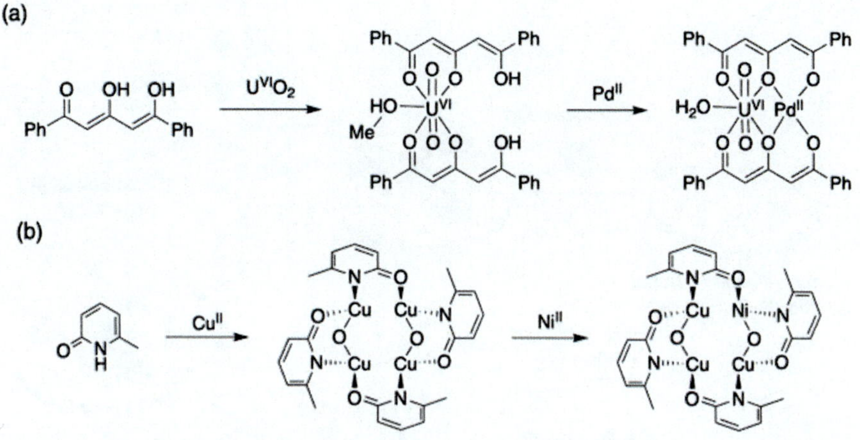

Fig. 2.3 Examples of syntheses of heterometallic complexes under kinetic control adopting **a** a metalloligand approach [18] and **b** a transmetalation approach [20]

of the desired heterometallic complex (Fig. 2.2b). A typical strategy is to use ligands and metals with orthogonal affinities based on their hard/soft characters [13]. In some other cases, difference between metals in the preferred coordination number/geometry [14] or ionic charge [15] is utilized.

However, the selective synthesis based on thermodynamic control sometimes fails, especially when the metals have similar thermodynamic properties, for example, in the cases of some transition metals or lanthanides. As discussed in Chap. 1, stepwise kinetic control is a good solution in such a case. As for the syntheses of heterometallic complexes, sequential addition of metals to each coordination site of a hetero-ditopic ligand is frequently used, which is called a metalloligand approach (Fig. 2.3a) [16–18]. On the other hand, site-selective transmetalation of a preformed homometallic complex is emerging in recent years as an alternative strategy to achieve kinetic control, especially when the metalloligand approach is not applicable (Fig. 2.3b) [19–24]. Still, it is sometimes difficult to kinetically differentiate the metal sites in homometallic complexes, resulting in a low yield or selectivity [25–27]. Therefore, it is desirable to develop a method of precisely controlling the transmetalation process for the synthesis of a heterometallic complex with high selectivity.

In this work, I employed redox switching of metal centers because it is well known to affect the properties of metal centers [28, 29]. From this viewpoint, I propose a stepwise strategy combining site-selective redox switching and transmetallation for kinetically controlled, highly selective synthesis of heterometallic complexes (Fig. 2.4). In this method, a ligand with inequivalent coordination sites, a redox-active metal, and another metal are used. Firstly, a homometallic complex of the redox-active metal is constructed by normal complexation with the ligands. Secondly, some of the chemically inequivalent metal centers are site-selectively oxidized. Thirdly, the non-oxidized metal centers are site-selectively transmetalated with another metal. Finally, the oxidized metal center(s) are reduced back to their original states. Among these

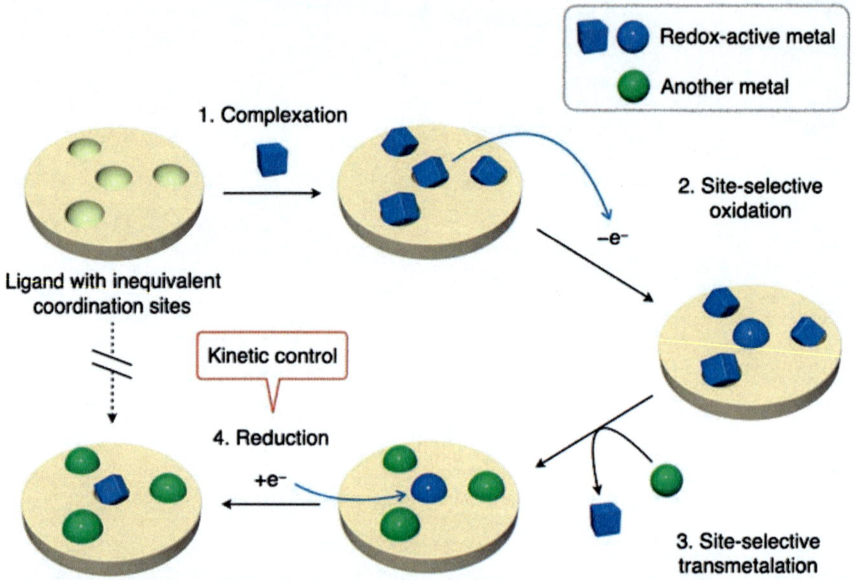

Fig. 2.4 Schematic of the strategy proposed in this research for highly selective synthesis of heterometallic complexes

four steps, the early three steps can be conducted under thermodynamic control. The last step needs to be conducted under kinetic control to give the desired complex selectively, and this can be guaranteed by kinetic stability of the polynuclear framework and by a relatively mild condition required for reduction.

The effectiveness of this strategy was demonstrated in the synthesis of a Co^{II}–Ni^{II} heterometallic complex (Fig. 2.5). I employed ligand **L1** as a ligand with inequivalent coordination sites, $Co^{II/III}$ as a redox-active metal, and Ni^{II} as another metal. Starting from ligand **L1**, the homometallic complex $[Co^{II}_4\mathbf{L1}_3X_6]$ (**1-Co^{II}_4**; X = a coordinated solvent molecule or counteranion), the site-selectively oxidized complex $[Co^{III}Co^{II}_3\mathbf{L1}_3X_6]$ (**2-$Co^{III}Co^{II}_3$**), the site-selectively transmetalated complex $[Co^{III}Ni^{II}_3\mathbf{L1}_3X_6]$ (**3-$Co^{III}Ni^{II}_3$**), and the finally, reduced complex $[Co^{II}Ni^{II}_3\mathbf{L1}_3X_6]$ (**4-$Co^{II}Ni^{II}_3$**) were successfully prepared stepwise with high selectivities and characterized well.

2.2 Step 1: Complexation with a Redox-Active Metal

As a ligand with inequivalent coordination sites, I employed ligand **L1** (Fig. 2.6). I previously found that ligand **L1** can self-assemble with Zn^{II} ions to form an unsymmetrical tetranuclear complex, $[Zn^{II}_4\mathbf{L1}_3X_6]$ (X = a coordinated solvent molecule or counteranion) [30]. In this complex, two kinds of chemically inequivalent metal

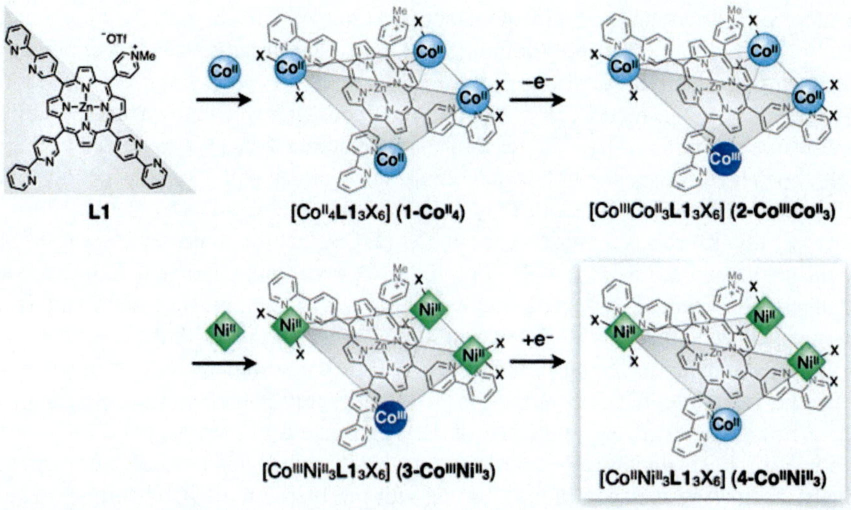

Fig. 2.5 Synthetic scheme of a Co^{II}–Ni^{II} heterometallic complex, [$Co^{II}Ni^{II}_3$**L1**$_3X_6$] (**4-**$\mathbf{Co^{II}Ni^{II}_3}$), demonstrated in this research using the aforementioned strategy

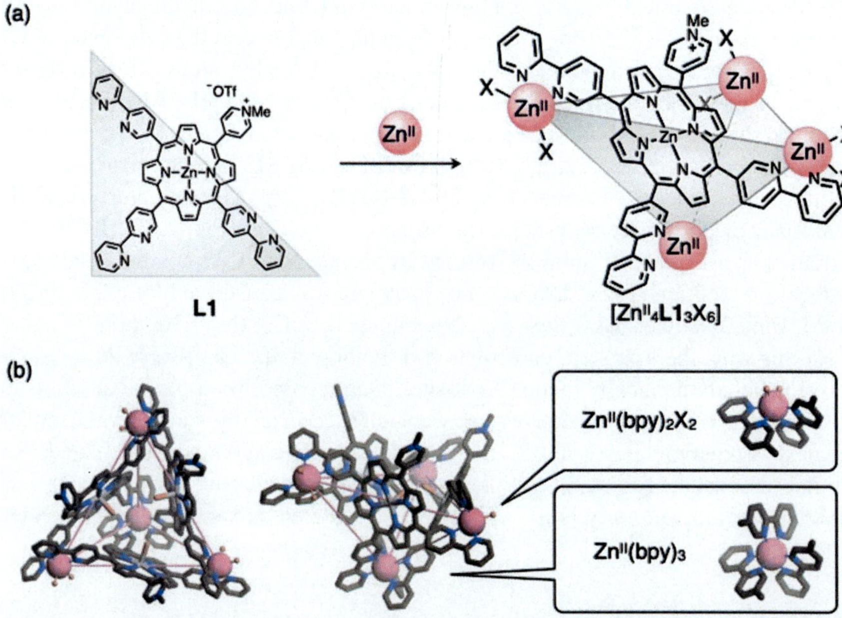

Fig. 2.6 Ligand **L1** and complex [Zn^{II}_4**L1**$_3X_6$] (X = a coordinated solvent molecule or counteranion) developed in a previous study [30]. **a** Reaction scheme of self-assembly of **L1** with Zn^{II} into [Zn^{II}_4**L1**$_3X_6$]. **b** Molecular structure of [Zn^{II}_4**L1**$_3X_6$] from top and side views. Insets show the partial structures around the chemically inequivalent metal centers, $Zn^{II}(bpy)_3$ and $Zn^{II}(bpy)_2X_2$

centers were arranged by the inequivalent coordination sites of **L1**: $Zn^{II}(bpy)_3$ and $Zn^{II}(bpy)_2X_2$ (bpy = 2,2′-bipyridine) (Fig. 2.6b). This structure seemed suitable to perform site-selective oxidation.

As a redox-active metal, $Co^{II/III}$ was chosen because it is well-known that the labile nature of Co^{II} dramatically changes into an inert nature when it was oxidized to Co^{III} [28, 29]. Therefore, as the first step, complexation of **L1** with Co^{II} was conducted. Treatment of **L1**·OTf with $Co(OTf)_2$ (1.33 equiv.) in CD_3CN/D_2O = 19:1 at 70 °C successfully afforded complex $[Co^{II}_4\mathbf{L1}_3X_6]$ (**1-Co**$^{II}_4$) as the major product in 65% yield (estimated from ^{1}H NMR) (Fig. 2.7a). A coordinated solvent molecule or counteranion denoted by X could not be determined in solution state due to the fast exchange. The yield was calculated as a sum of every species with different Xs.

The paramagnetic ^{1}H NMR spectrum at 343 K showed three sets of signals with the 1:1:1 integral ratio in the 30–110 ppm region, which can be assigned to three bpy moieties of one molecule of **L1** coordinated to paramagnetic Co^{II} ions (Fig. 2.7b). Each signal was assigned to 6,6′-, 3,3′-, and 5-positions of bpy, respectively, comparing the spectrum at 300 K with the literature of $[Co^{II}(bpy)_3]^{2+}$ [31, 32], because the through-bond proximity to paramagnetic metal atom is known to dominate downfield shifting [33–35]. Some signals of hydrogens at the 5-, 6-, and 6′-positions were broadened to different degrees, probably due to the dynamic exchange of X. The two chemically equivalent bpy moieties in **L1** became inequivalent because the connection to the $Co^{II}(bpy)_3$ moiety (C_3 symmetry) lowers the symmetry of **L1** (C_{2v}) to C_1. In the diamagnetic region, the signal of methyl protons was assigned by its characteristic high intensity derived from 3 protons and relatively non-shifted chemical shift around 5 ppm. The presence of only one set of signals for **L1** is consistent with the C_3-symmetric structure of **1-Co**$^{II}_4$. The ESI-MS measurement predominantly showed a series of signals for $[Co^{II}_4\mathbf{L1}_3(OTf)_{11-n}]^{n+}$. The single-crystal XRD analysis showed the detailed molecular structure of **1-Co**$^{II}_4$ (Fig. 2.7c). **1-Co**$^{II}_4$ had a triangular-pyramidal framework formed by inequivalent Co^{II} centers, analogous to the aforementioned $[Zn^{II}_4\mathbf{L1}_3X_6]$. The apex site was constituted by a $Co^{II}(bpy)_3$ motif, while the other basal sites were constituted by a $Co^{II}(bpy)_2X_2$ motif.

In this way, the first step, construction of homometallic complex with a redox-active metal at chemically inequivalent sites, was accomplished. As all attempts to obtain pure **1-Co**$^{II}_4$ in a preparative scale were not successful, the sample was used for the next step without purification, and the isolated yield could not be calculated. For similar reasons, the following complexes were also used without purification unless otherwise noted, and only NMR yields were calculated as a sum of every species with different Xs.

2.3 Step 2: Site-Selective Oxidation

Next, site-selective oxidation of chemically inequivalent metal centers was conducted. In **1-Co**$^{II}_4$, the $Co^{II}(bpy)_3$ site is expected to be more easily oxidized than $Co^{II}(bpy)_2X_2$ sites in thermodynamic equilibrium because the electron donation

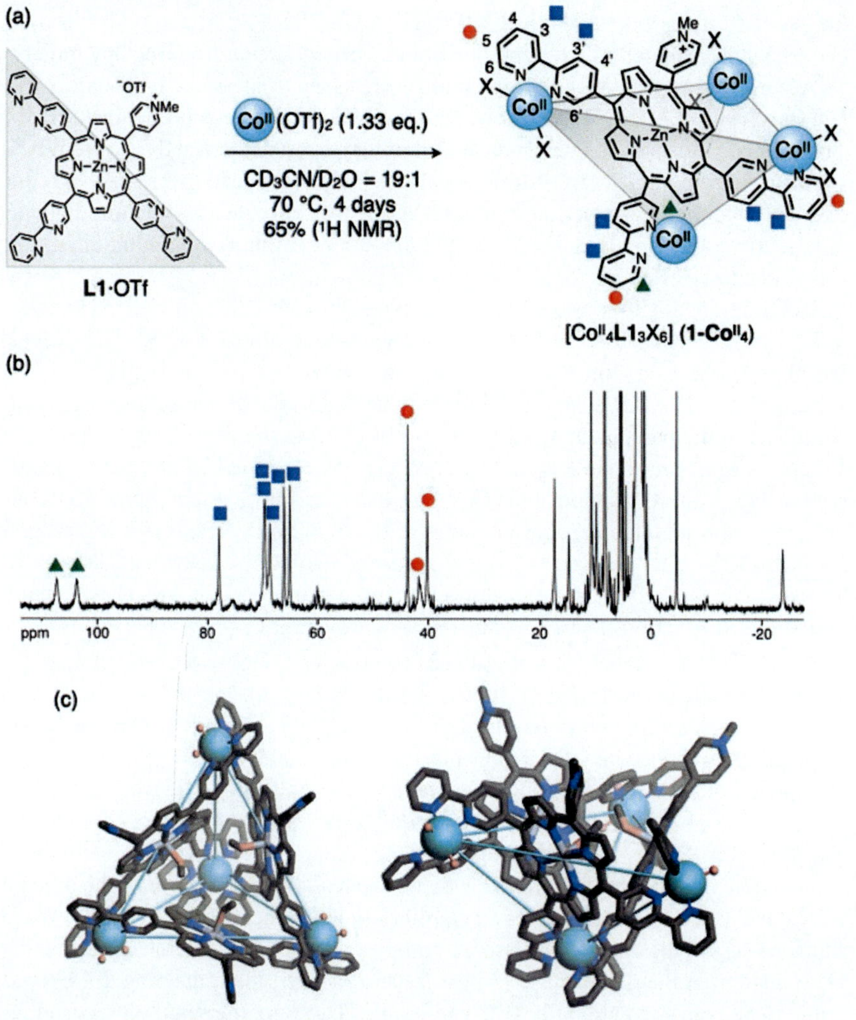

Fig. 2.7 Preparation and characterization of complex $[Co^{II}_4L1_3X_6]$ (**1-Co$^{II}_4$**). **a** Reaction scheme. The numbering of positions is shown in one of the bpy moieties. **b** Paramagnetic ^{1}H NMR spectrum (CD_3CN/D_2O = 19:1, 343 K, 500 MHz). **c** Single-crystal XRD structure of a MeOH adduct in top and side views. Hydrogen atoms, disorder, non-coordinated anions, and solvent molecules are omitted for clarity. Color code: Co^{II}, cyan; Zn, gray; C, black; N, blue; O, red. Lines are drawn between Co centers to guide the eye

from bpy ligands should lower the redox potential of $Co^{III/II}$. Attempts to electrochemically determine the redox potential of each Co^{II} site of **1-Co$^{II}_4$** failed because of high irreversibility of these redox couples and interference of redox reaction at the porphyrin core of **L1** at high potentials. Nonetheless, the literature values of the standard redox potentials of [$Co^{III/II}(OH_2)_6$] (+1.92 V) [36] and [$Co^{III/II}(bpy)_3$] (+0.32 V) [37] supported the hypothesis of electron donation from the bpy moiety.

As an initial screening, various oxidants were tested to site-selectively oxidize the $Co^{II}(bpy)_3$ site of **1-Co$^{II}_4$**. H_2O_2 showed extensive decomposition probably due to too strong oxidization power. DDQ indicated simultaneous oxidation at the $Co^{II}(bpy)_2X_2$ site with coordination of OH^- formed by the basicity of reduced species. $AgNO_3$ and 1,8-naphthoquinone showed no redox reactions probably due to a thermodynamic or kinetic reason. Finally, a stoichiometric amount of cerium(IV) ammonium nitrate was found to be effective for site-selective oxidation.

Accordingly, **1-Co$^{II}_4$** was treated with cerium(IV) ammonium nitrate (0.333 equiv. to **L1**) in CD_3CN/D_2O = 19:1 at room temperature to afford [$Co^{III}Co^{II}_3$**L1**$_3X_6$] (**2-Co^{III}Co$^{II}_3$**), where only the $Co^{II}(bpy)_3$ site was oxidized to $Co^{III}(bpy)_3$ (Fig. 2.8a). A paramagnetic ^{1}H NMR spectrum of the product showed that one set of bpy signals shifted from the paramagnetic region (30–110 ppm) to the diamagnetic region (7–11 ppm), while the other two sets of bpy signals remained in the paramagnetic region (Fig. 2.8b). Oxidation at the $Co^{II}(bpy)_3$ site can only explain such shift of one bpy signal sets maintaining the symmetry of the framework. The ESI-MS measurement predominantly showed the signal series of [$Co^{III}Co^{II}_3$**L1**$_3(NO_3)(OTf)_{11-n}]^{n+}$. The single-crystal XRD analysis showed that the framework of **1-Co$^{II}_4$** was maintained (Fig. 2.8c). In comparison with **1-Co$^{II}_4$**, the average Co–N bond distance at the $Co(bpy)_3$ site was significantly shorter (2.002(4) Å vs. 2.046(6)Å), while that at the $Co(bpy)_2X_2$ sites was similar (2.106(6) Å vs. 2.122(6) Å). In the literature, Co–N bond distances are known to be shorter in [$Co^{III}(bpy)_3$] than in [$Co^{II}(bpy)_3$] (1.938 Å vs. 2.128 Å as median values in Cambridge Structural Database [38]). Thus, this contraction of bond distances suggests that only the $Co^{II}(bpy)_3$ site was oxidized.

In this way, oxidation of **1-Co$^{II}_4$** into **2-Co^{III}Co$^{II}_3$** was site-selectively completed. The yield was 83% as estimated from ^{1}H NMR. Although 14% of **1-Co$^{II}_4$** remained, the reaction yield was not improved even when an additional amount of the oxidant was added. It seems that there was an equilibrium with other oxidized species. When a large excess of the oxidant was used, a complex mixture was generated presumably via oxidation at the $Co^{II}(bpy)_2X_2$ sites. As shown later, the remaining **1-Co$^{II}_4$** was found to be converted similarly to **2-Co^{III}Co^{II}**. Therefore, the sample was used for the next step without further treatment.

2.4 Step 3: Site-Selective Transmetalation

Next, site-selective transmetalation at the unoxidized $Co^{II}(bpy)_2X_2$ sites was conducted. In **2-Co^{III}Co$^{II}_3$**, the $Co^{III}(bpy)_3$ site is expected to be more stable against transmetalation than the $Co^{II}(bpy)_2X_2$ sites. From a thermodynamic point of view, the affinity of bpy to Co^{III} is much higher than that to Co^{II} as can be seen in the formation constant of [$Co(bpy)_3$] complexes ($10^{43.0}$ for Co^{III} vs. $10^{15.9}$ for Co^{II} in H_2O, as calculated from the redox potential of [$Co^{III/II}(OH_2)_6$] [36] and [$Co^{III/II}(bpy)_3$]

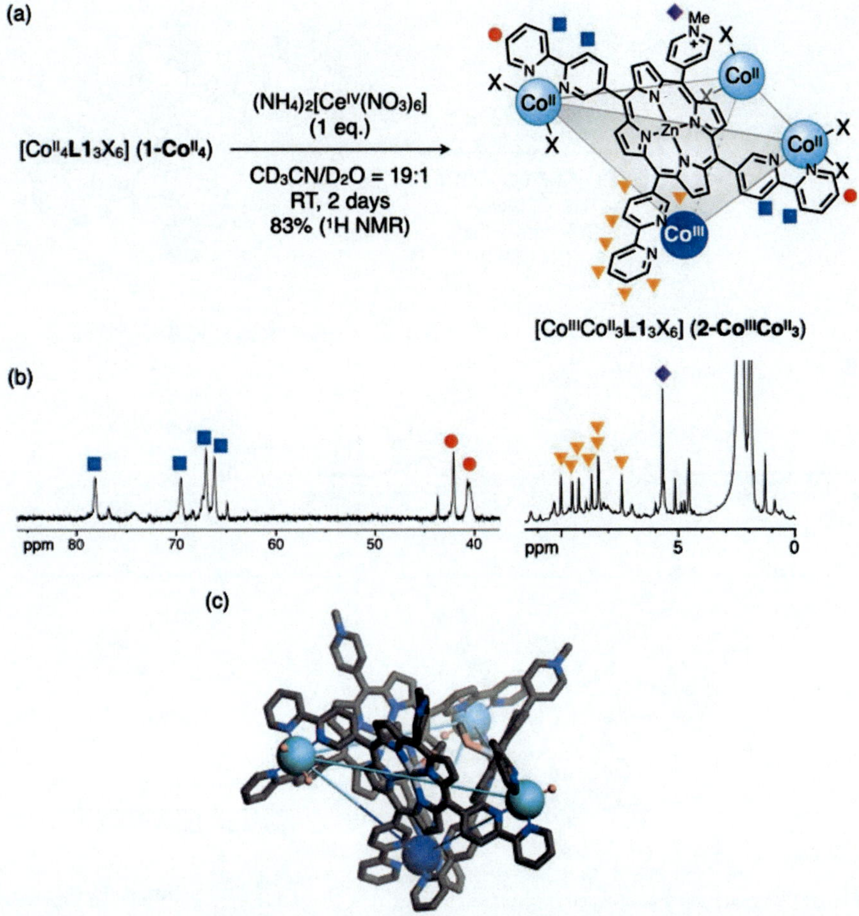

Fig. 2.8 Preparation and characterization of complex $[Co^{III}Co^{II}_3\mathbf{L1}_3X_6]$ (**2-Co^{III}Co$^{II}_3$**). **a** Reaction scheme. **b** Paramagnetic ^{1}H NMR spectrum (CD_3CN/D_2O = 19:1, 343 K, 500 MHz). **c** Single-crystal XRD structure of a MeOH adduct. Hydrogen atoms, disorder, non-coordinated anions, and solvent molecules are omitted for clarity. Color code: Co^{III}, dark blue; Co^{II}, cyan; Zn, gray; C, black; N, blue; O, red. Lines are drawn between Co centers to guide the eye

[37] and the formation constant of $[Co^{II}(bpy)_3]$ [39]). From a kinetic point of view, octahedral Co^{III} complexes are known to be kinetically much more inert than Co^{II} ones [28, 29], and the $Co(bpy)_3$ site has more bridging to the other Co centers than the $Co(bpy)_2X_2$ sites. I selected Ni^{II} as another metal to introduce because the thermodynamic affinity of bpy to Ni^{II} is intermediate between Co^{III} and Co^{II} (the formation constant of $[Ni(bpy)_3]$ is $10^{20.2}$ in H_2O) [39]. Accordingly, **2-Co^{III}Co$^{II}_3$** was treated with $Ni(OTf)_2$ (2 equiv. to **L1**) in CD_3CN/D_2O = 19:1 at 70°C to afford $[Co^{III}Ni^{II}_3\mathbf{L1}_3X_6]$ (**3-Co^{III}Ni$^{II}_3$**) (Fig. 2.9a).

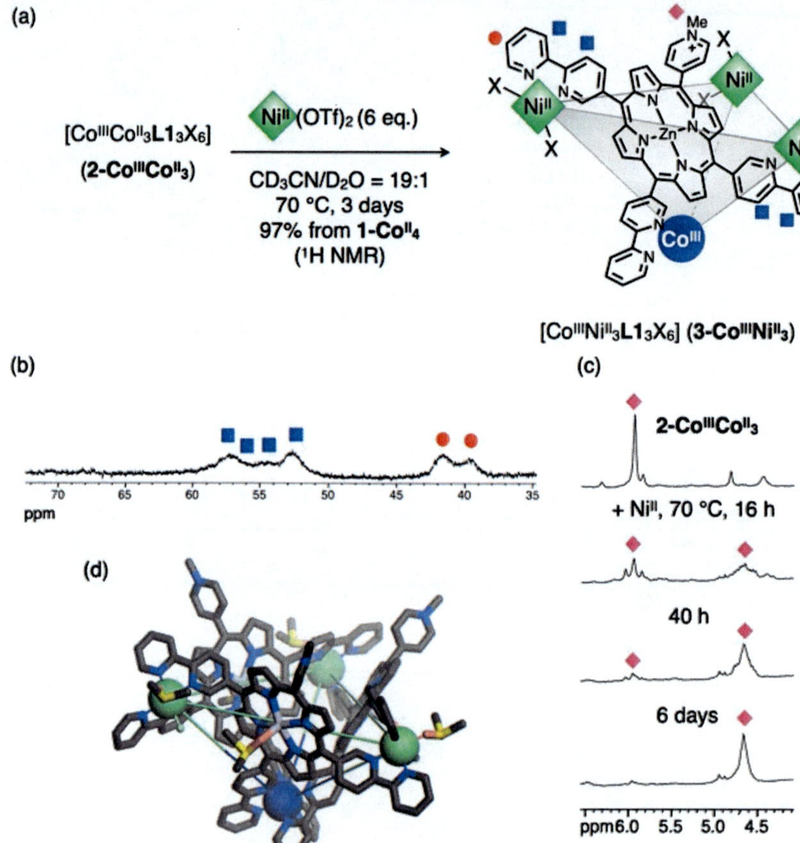

Fig. 2.9 Preparation and characterization of complex $[Co^{III}Ni^{II}_3\mathbf{L1}_3X_6]$ ($\mathbf{3\text{-}Co^{III}Ni^{II}_3}$). **a** Reaction scheme. **b** Paramagnetic ^1H NMR spectrum (CD_3CN/D_2O = 19:1, 343 K, 500 MHz). **c** Time-course ^1H NMR spectra (CD_3CN/D_2O = 19:1, 300 K, 500 MHz). **d** Single-crystal XRD structure of an adduct with DMSO and chloride. Hydrogen atoms, disorder, non-coordinated anions, and solvent molecules are omitted for clarity. Color code: Co^{III}, dark blue; Ni^{II}, yellow-green; Zn, gray; Cl, light yellow-green; C, black; N, blue; O, red. Lines are drawn between the Co and Ni centers to guide the eye

The paramagnetic ^1H NMR spectrum showed the disappearance of the $Co^{II}(bpy)_2X_2$ signals and appearance of new broad signals with different chemical shifts, typical for Ni^{II}–bpy complexes (Fig. 2.9b) [32]. In the diamagnetic region, the signal of methyl group of **L1** initially changed into a complicated pattern and then converged to one signal (Fig. 2.9c). These results indicate that all the three $Co^{II}(bpy)_2X_2$ sites were stepwise transmetalated with Ni^{II}. The ESI-MS measurement showed the signal series of $[Co^{III}Ni^{II}_3\mathbf{L}_3(NO_3)(OTf)_{11-n}]^{n+}$, supporting that transmetalation did not occur at the $Co^{III}(bpy)_3$ site. The XPS measurement after removal of free metal ions by reprecipitation showed that the atomic ratio of Co:Ni was 1:3.1. The single-crystal XRD analysis showed that the framework of **2-Co^{III}Co$^{II}_3$** was maintained (Fig. 2.9d). The average Co–N bond distance at the Co^{III}(bpy)$_3$ site was 1.936(3) Å, and the average Ni–N bond distance at the Ni^{II}(bpy)$_2$X$_2$ sites was 2.056(4) Å. These values agree well with typical values of Co^{III} (1.938 Å) and Ni^{II} (2.086 Å), but not with Co^{II} (2.128 Å as median values of [M(bpy)$_3$] in Cambridge Structural Database) [38]. This crystal was obtained as an adduct of chloride ions, which probably came from decomposition of CDCl$_3$ used for crystallization [40].

In this way, transmetalation of **2-Co^{III}Co$^{II}_3$** into **3-Co^{III}Ni$^{II}_3$** was site-selectively completed. The yield was 97% from **1-Co$^{II}_4$** as estimated from ^{1}H NMR. This yield indicates the remaining **1-Co$^{II}_4$** (14%) in the sample of **2-Co^{III}Co$^{II}_3$** was also converted to **3-Co^{III}Ni$^{II}_3$**, probably via oxidation by other oxidized species.

2.5 Step 4: Reduction

As the final step, reduction of the initially oxidized metal center was conducted. Since the thermodynamic affinity of bpy to Ni^{II} is higher than Co^{II}, the reduction product of **3-Co^{III}Co$^{II}_3$** would be thermodynamically unstable against metal exchange between the Co^{II}(bpy)$_3$ site and the Ni^{II}(bpy)$_2$X$_2$ sites. Nonetheless, considering the kinetic stability of the polynuclear framework consisting of multiple coordination bonds, it may be possible to conduct reduction under kinetic control without causing metal exchange. From this viewpoint, **3-Co^{III}Co$^{II}_3$** was treated with ^{n}Bu$_4$N·I (1.5 equiv. to **L1**) in CD$_3$CN/D$_2$O = 19:1 at 70 °C to afford [Co^{II}Ni$^{II}_3$**L1**$_3$X$_6$] (**4-Co^{II}Ni$^{II}_3$**) (Fig. 2.10a). ^{n}Bu$_4$N·I was chosen as a reductant because it is known to be effective for similar complexes [29].

The paramagnetic ^{1}H NMR spectrum showed the re-appearance of one set of sharp Co^{II}–bpy signals, while the signals of the Ni^{II}(bpy)$_2$X$_2$ sites were unchanged (Fig. 2.10b). In the diamagnetic region, the methyl signal of **3-Co^{III}Co$^{II}_3$** disappeared and a new signal appeared. This result indicates that the Co^{III} center was reduced to Co^{II} without affecting the other parts of the structure. The ESI-MS measurement showed a series of signals for [Co^{II}Ni$^{II}_3$**L1**$_3$(NO$_3$)$_2$(OTf)$_{9-n}$]$^{n+}$. The UV-Vis absorption spectrum of **4-Co^{II}Ni$^{II}_3$** was similar to that of **1-Co$^{II}_4$** but markedly different from those of **2-Co^{III}Co$^{II}_3$** and **3-Co^{III}Ni$^{II}_3$**, suggesting the absence of Co^{III} character (Fig. 2.10c).

In this way, reduction of **3-Co^{III}Ni$^{II}_3$** to **4-Co^{II}Ni$^{II}_3$** was successfully confirmed. However, the yield in an initial attempt was 48% as estimated from ^{1}H NMR. This result suggests that the reaction was not completely under kinetic control, and that decomposition of the product occurred in part. To prevent possible decomposition, the sample of **3-Co^{III}Ni$^{II}_3$** was partially purified by reprecipitation from MeCN/H$_2$O = 1:2 with NH$_4$PF$_6$ (50 equiv. to **L1**), and the amount of the reductant was decreased to 0.417 equiv. to **L1**. In this case, the yield was improved to 95% as estimated from ^{1}H NMR, which means reduction surely proceeded under kinetic control.

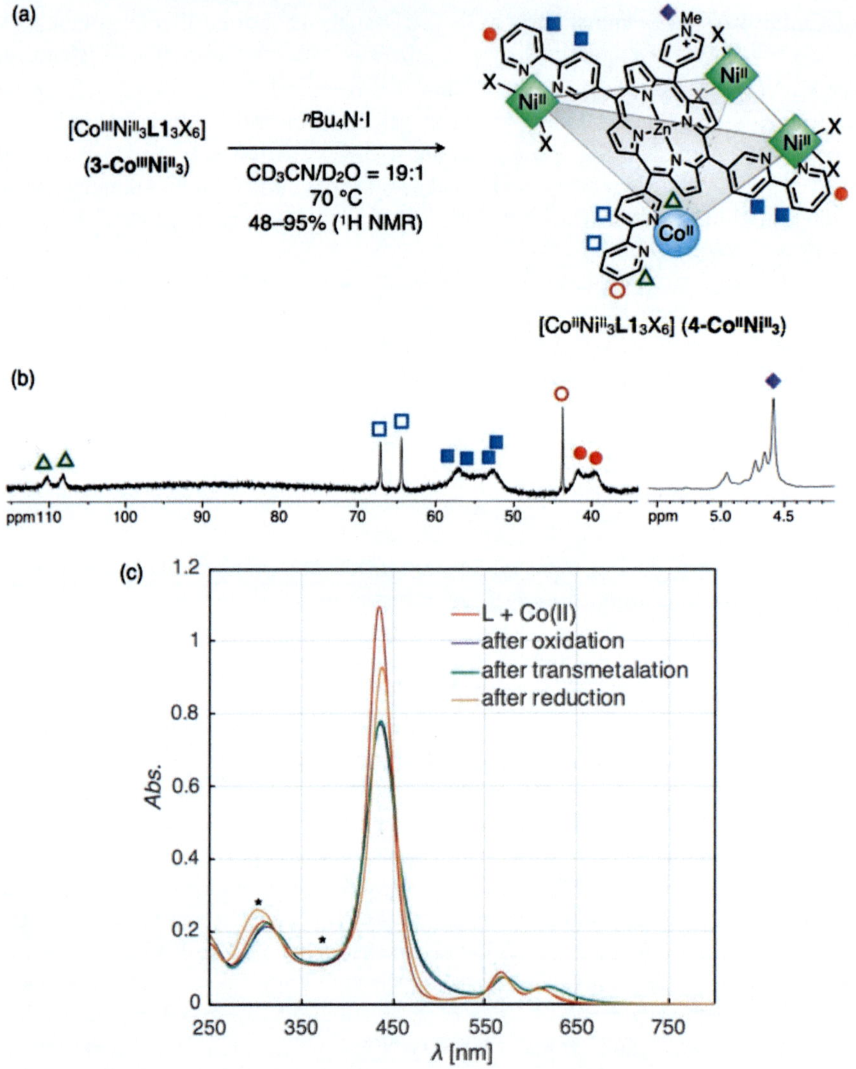

Fig. 2.10 Preparation and characterization of complex $[Co^{II}Ni^{II}_3\mathbf{L1}_3X_6]$ (**4-$Co^{II}Ni^{II}_3$**). **a** Reaction scheme. **b** Paramagnetic ^{1}H NMR spectrum (CD_3CN/D_2O = 19:1, 343 K, 500 MHz). **c** UV-Vis absorption spectra (MeCN, 298 K, $l = 0.10$ cm). Each sample prepared in CD_3CN/D_2O = 19:1 was diluted to 25 times with CH_3CN. The total concentration of **L** was 40 μM. The asterisks denote the absorption bands of triiodide ion

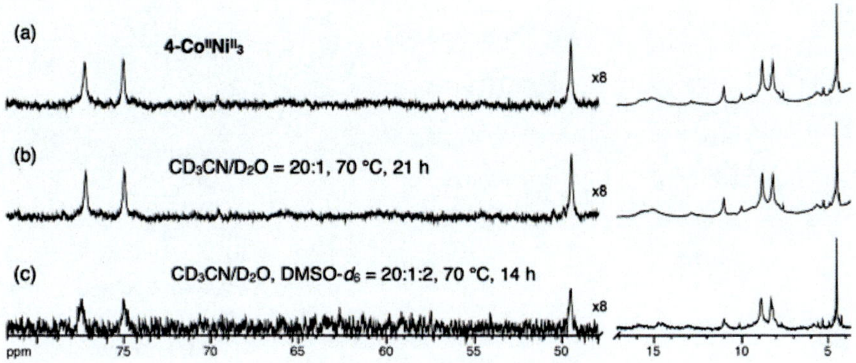

Fig. 2.11 Stability test of **4-Co^{II}Ni$^{II}_3$**. Paramagnetic ^{1}H NMR spectrum (**a**, **b** $CD_3CN/D_2O = 20:1$; **c** $CD_3CN/D_2O/DMSO\text{-}d_6 = 20:1:2$, 300 K, 500 MHz)

Overall, the heterometallic complex **4-Co^{II}Ni$^{II}_3$** was selectively obtained in 92% yield from **1-Co$^{II}_4$** via stepwise site-selective redox switching and transmetalation.

Surprisingly, **4-Co^{II}Ni$^{II}_3$** prepared by the modified procedure showed no sign of decomposition when heated at 70 °C in $CD_3CN/D_2O = 20:1$ for 21 h, or even in $CD_3CN/D_2O/DMSO\text{-}d_6 = 20:1:2$ for 24 h (Fig. 2.11). This high kinetic stability would come from the polynuclear framework consisting of multiple coordination bonds.

2.6 Comparison with Other Strategies

To confirm the effectiveness of the strategy used here, I conducted three control experiments using other strategies.

Firstly, the one-step strategy was examined. **L1**·OTf was treated with Co^{II}(OTf)$_2$ (0.33 equiv.) and Ni^{II}(OTf)$_2$ (1 equiv.) simultaneously in $CD_3CN/D_2O = 19:1$ at 70 °C (Fig. 2.12a). The paramagnetic ^{1}H NMR spectrum showed only signals of Ni^{II}–bpy but no discernable signals of Co^{II}–bpy (Fig. 2.12b). In the region of methyl protons, many weak signals were observed. This result indicates non-selective formation of Ni^{II}-containing complexes. This failure would be caused by the higher thermodynamic stability of a Ni^{II}(bpy)$_3$ motif than Co^{II}(bpy)$_3$. The one-step strategy could not provide the kinetic control necessary to afford **4-Co^{II}Ni$^{II}_3$** selectively.

Secondly, the transmetalation strategy without redox switching was examined. **1-Co$^{II}_4$** was treated with Ni(OTf)$_2$ (2 equiv. to **L1**) in $CD_3CN/D_2O = 19:1$ at 70 °C (Fig. 2.13a). Similarly to the previous experiment, the paramagnetic ^{1}H NMR spectrum showed only signals of Ni^{II}–bpy but no discernable signals of Co^{II}–bpy (Fig. 2.13b). In the region of methyl protons, some weak signals were observed. This result indicates transmetalation at the Co^{II}(bpy)$_3$ site and non-selective formation of Ni^{II}-containing complexes. This failure would be caused by the insufficient

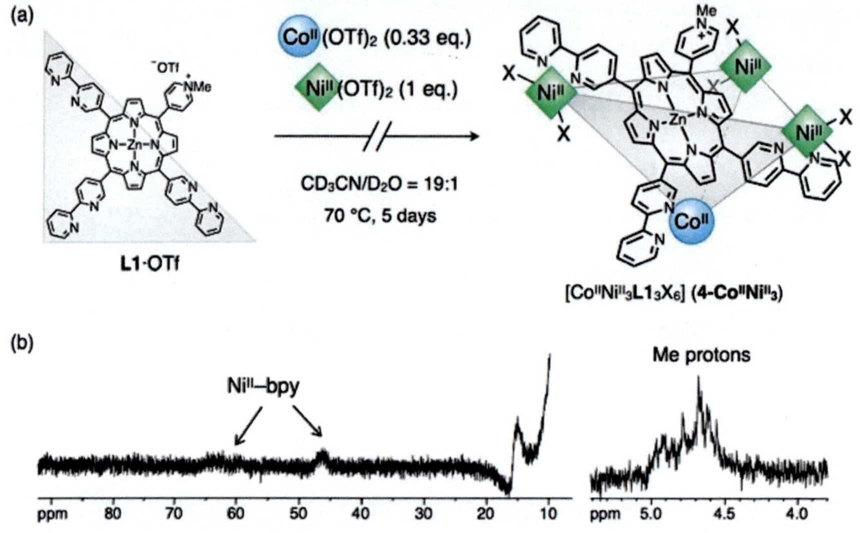

Fig. 2.12 Control experiment of the one-step strategy. **a** Reaction scheme. **b** Paramagnetic ^1H NMR spectrum (CD_3CN/D_2O = 19:1, 300 K, 500 MHz)

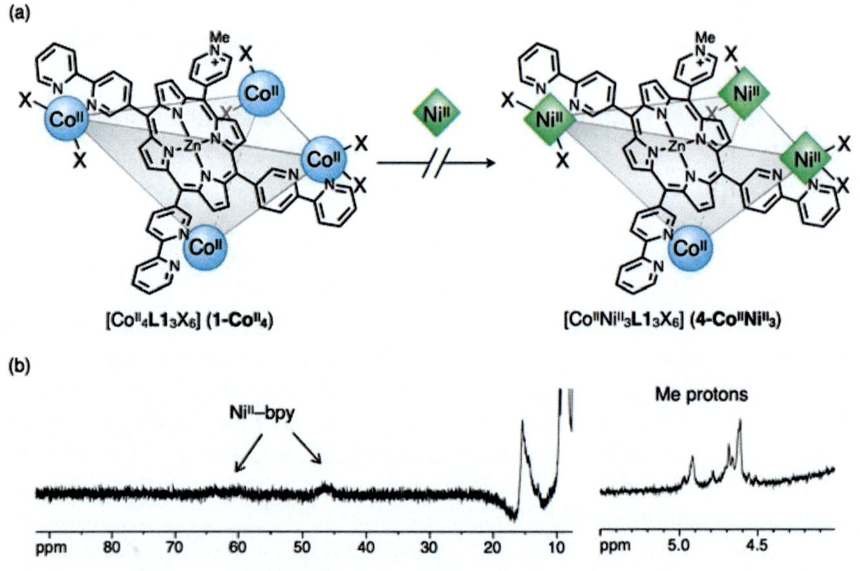

Fig. 2.13 Control experiment of the transmetalation strategy without redox switching. **a** Reaction scheme. **b** Paramagnetic ^1H NMR spectrum (CD_3CN/D_2O = 19:1, 300 K, 500 MHz)

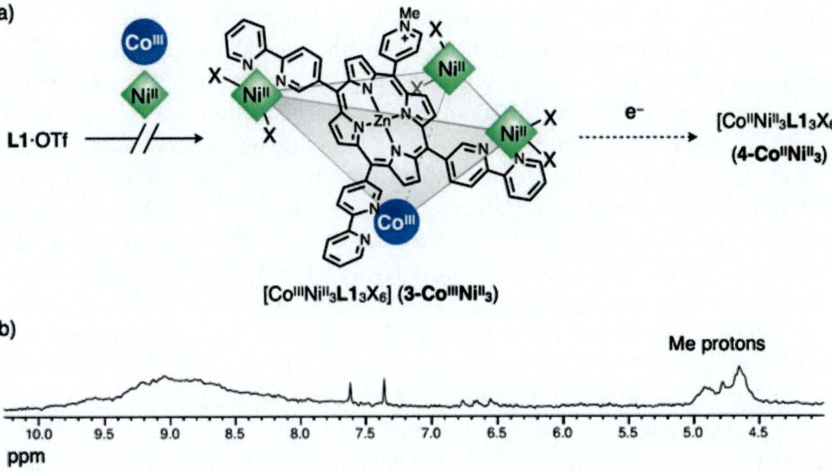

Fig. 2.14 Control experiment of the redox-switching strategy without transmetalation. **a** Reaction scheme. **b** ^{1}H NMR spectrum (CD_3CN/D_2O = 19:1, 300 K, 500 MHz)

kinetic control. The transmetalation condition was harsher than that of reduction, because a longer reaction time was necessary, and because a high concentration of free metal ions was present in solution. Although the transmetalation strategy works in some cases [19–24], combination with redox switching was necessary in the case of **4-Co^{II}Ni$^{II}_3$**.

Thirdly, the redox-switching strategy without transmetalation was examined. Since **3-Co^{III}Ni$^{II}_3$** is expected to be thermodynamically stable, direct synthesis of **3-Co^{III}Ni$^{II}_3$** from **L1** and subsequent reduction might be another route to **4-Co^{II}Ni$^{II}_3$**. Accordingly, **L1**·OTf was treated with $Co^{II}(OTf)_2$ (0.33 equiv.) and $Ni^{II}(OTf)_2$ (1 equiv.) in CD_3CN/D_2O = 19:1 at 70 °C, and then, $(NH_4)_2[Ce^{IV}(NO_3)_6]$ (0.33 equiv.) was added to the reaction mixture at 70 °C (Fig. 2.14a). In situ oxidation of Co^{II} was employed here because preparation of $Co^{III}(OTf)_3$ is not known. The ^{1}H NMR spectrum showed many weak signals, indicating non-selective formation of some complexes (Fig. 2.14b). This failure would be caused either by the high kinetic barrier to ligand exchange at a Co^{III} center or by the thermodynamic instability of **3-Co^{III}Ni$^{II}_3$** due to the smaller size of Co^{III} than that of Co^{II}. It seems that redox switching and introduction of another metal should be separately conducted to avoid complexity.

2.7 Conclusions

In this chapter, I have proposed a novel kinetically controlled, stepwise strategy for highly selective synthesis of heterometallic complexes. This strategy uses a redox-active metal and a ligand with inequivalent coordination sites and consists of four steps: complexation with a redox-active metal, site-selective oxidation, site-selective transmetalation, and reduction. In this strategy, the early three steps proceed under thermodynamic control, and the fourth step needs kinetic control. This kinetic control can be provided by the relatively mild condition of reduction and the high kinetic stability of polynuclear framework consisting of multiple coordination bonds.

The effectiveness of this strategy was demonstrated in the synthesis of a new heterometallic complex, **4-$Co^{II}Ni^{II}_3$**. In this synthesis, $Co^{II/III}$ worked as a redox-active metal whose affinity to ligand was dramatically increased upon oxidation. Tris-bipyridine ligand **L1** was employed to construct an $[M_4\mathbf{L1}_3X_6]$ framework with inequivalent metal sites. Ligand **L1** was treated with Co^{II} to yield **1-Co^{II}_4**, and then, it was site-selectively oxidized to give **2-$Co^{III}Co^{II}_3$**, and it was site-selectively transmetalated to give **3-$Co^{III}Ni^{II}_3$**, and finally, it was reduced to give **4-$Co^{II}Ni^{II}_3$** selectively. In contrast, this complex could not be selectively obtained either by the conventional one-step strategy or by the simple transmetalation. Therefore, this "site-selective redox switching and transmetalation" strategy is an excellent way to construct novel heterometallic complexes, which can exhibit functions specific to the combination of metals.

As an extension of this strategy, it may be possible to synthesize a hetero-trimetallic complex. As discussed in Sects. 2.2–2.4, the introduction of Ni^{II} ions was observed in a stepwise manner. This may allow introduction of less Ni^{II} ions and subsequent transmetalation with another metal ion. Meanwhile, complex **4-$Co^{II}Ni^{II}_3$** may be useful as a catalyst by binding a substrate at the Ni^{II} site and providing electron transfer from the Co^{II} site.

2.8 Experimental Section

2.8.1 Materials and Methods

Unless otherwise noted, solvents and reagents were purchased from TCI Co., Ltd., FUJIFILM Wako Pure Chemical Corporation Ltd., Kanto Chemical Co., and Sigma-Aldrich Co., and used without further purification.

Elemental analysis was conducted in the Microanalytical Laboratory, Department of Chemistry, School of Science, the University of Tokyo.

^{1}H, ^{13}C, ^{19}F, and 2D NMR spectra were recorded on a Bruker AVANCE III-500 (500 MHz) spectrometer. Wide-sweep paramagnetic NMR spectra were recorded with a spectral width (SW) of 200 ppm, a transmitter frequency offset (O1P) of 50.00 ppm, and a line width of 10.0 Hz unless otherwise noted. NMR yield estimation using an internal standard was conducted in a normal measurement method for the diamagnetic region. Tetramethylsilane was used as an internal standard (δ 0 ppm) for ^{1}H and ^{13}C NMR measurements when $CDCl_3$ was used as a solvent. A residual solvent signal was used for the calibration of ^{1}H NMR measurements when CD_3CN (δ 1.94 ppm), DMSO-d_6 (δ 2.50 ppm), or mixed solvents with them were used as a solvent. No corrections were conducted for ^{19}F NMR measurements. Abbreviations: s, singlet; d, doublet; t: triplet; br, broad; m, multiplet.

ESI-TOF MS data were recorded on a Micromass LCT Premier XE mass spectrometer. Unless otherwise noted, experimental conditions were as follows (Ion mode, positive; Desolvation temperature, 150 °C; Source temperature, 80 °C).

UV–Vis spectra were recorded on a JASCO V-770 UV–vis–NIR spectrophotometer. The experimental conditions were as follows (0.10 cm glass cell; λ = 250–800 nm; scanning rate, 100 nm/min; data acquisition intervals, 0.5 nm).

Single-crystal X-ray crystallographic analyses were performed using a Rigaku XtaLAB PRO MM007DW PILATUS diffractometer, and obtained data were processed by using CrysAlisPro 1.171.39.7e (Rigaku OD, 2015) software and analyzed by Olex [2] 1.2.10 (OlexSys Ltd., 2018) software [41] and SHELXL-2017/1 software [42]. Crystallographic data in this chapter can be obtained free of charge from the Cambridge Crystallographic Data Centre (http://www.ccdc.cam.ac.uk/data_request/cif).

XPS spectra were recorded using an Ulvac-Phi PHI 5000 VersaProbe III spectrometer and an Al Kα X-ray source ($h\nu$ = 1486.6 eV).

2.8.2 *Synthesis of Compounds*

2.8.2.1 Synthetic Route to L1·OTf·3H_2O

L1·OTf·3H_2O was synthesized via the route below according to my master thesis (Scheme 2.1) [30].

5-Methyl-2,2′-Bipyridine (**5**)

This compound was prepared according to a previous report [43]. The ^{1}H NMR data of the compound were identical with those reported [43].

^{1}H NMR ($CDCl_3$, 300 K, 500 MHz): δ 8.66 (ddd, J = 4.8, 1.8, 0.9 Hz, 1H), 8.50 (dt, J = 2.2, 0.7 Hz, 1H), 8.36 (dt, J = 8.0, 1.1 Hz, 1H), 8.28 (dd, J = 8.1, 0.3 Hz, 1H), 7.79 (ddd, J = 8.0, 7.5, 1.8 Hz, 1H), 7.62 (dd, J = 8.1, 1.7 Hz, 1H), 7.27 (ddd, J = 7.4, 4.9, 1.2 Hz, 1H), 2.38 (s, 1H) ppm.

5-Formyl-2,2′-Bipyridine (**6**)

This compound was prepared by a procedure modified from a previous report [43]. A three-necked 1 L flask was equipped with a reflux condenser and a magnetic stirring bar, and to the flask were added 5-methyl-2,2′-bipyridine (**5**) (7.50 g, 44.1 mmol, 1 equiv.), NBS (19.6 g, 110 mmol, 2.49 equiv.) and CCl_4 (600 mL). The mixture was degassed and bubbled with Ar gas. To the mixture was added AIBN (0.867 g,

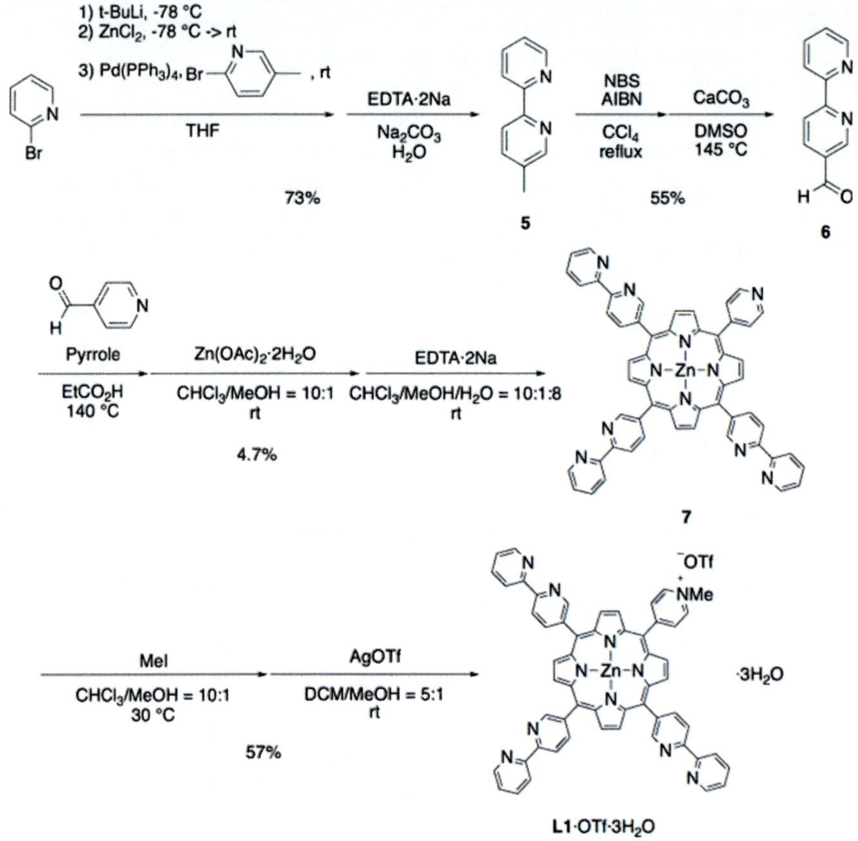

Scheme 2.1 Synthetic route of **L1**·OTf·$3H_2O$

5.28 mmol, 12.0 mol%). The reaction mixture was degassed, and the flask was replaced with Ar gas. The reaction mixture was stirred and heated at reflux for 6 h. The resulting precipitate was removed with suction and washed with CCl_4. The filtrate and wash liquid were combined and evaporated to afford a brown solid. The solid was dissolved in CH_2Cl_2 (200 mL), and the solution was washed with Na_2CO_3 aqueous solution (pH 9, 200 mL $\times$ 2). The organic layer was evaporated to afford a brown solid. The solid was dissolved in Et_2O (100 mL), and the solution was filtered through SiO_2 short-path column with suction. The filtrate was evaporated to afford a yellow solid (12.5 g) containing 5-bromomethyl-2,2′-bipyridine, 5-dibromomethyl-2,2′-bipyridine as a desired intermediate and 5-tribromomethyl-2,2′-bipyridine.

A 1 L three-necked flask was equipped with a reflux condenser and a magnetic stirring bar, and to the flask were added the resultant solid (12.5 g), DMSO (370 mL), and $CaCO_3$ (15.6 g, 155 mmol, 2.1 equiv. to Br atom). The reaction mixture was degassed and bubbled with Ar gas. The reaction mixture was stirred at 145 °C for 4.5 h. To the reaction mixture was added H_2O (1.85 L) and Et_2O (400 mL). The mixture was filtered with suction and washed with Et_2O. The organic layer of the filtrate was separated. The aqueous layer was extracted with Et_2O (400 mL × 4). The organic layers were combined, evaporated, and the resultant solid was dissolved in Et_2O (150 mL). The solution was washed with brine, dried over Na_2SO_4, and evaporated to afford a pale-yellow solid containing 5-formyl-2,2′-bipyridine (**6**) as a main product. Recrystallization of the solid from hot *n*-hexane gave **6** as a pale-yellow solid (4.43 g, 24.0 mmol, 55%). The ^{1}H NMR data were identical with those reported [43].

^{1}H NMR ($CDCl_3$, 300 K, 500 MHz): δ 10.18 (s, 1H), 9.13 (dd, $J = 2.1, 0.8$ Hz, 1H), 8.73 (ddd, $J = 4.8, 1.8, 0.9$ Hz, 1H), 8.62 (dd, $J = 8.2, 0.6$ Hz, 1H), 8.52 (dt, $J = 8.0, 1.1$ Hz, 1H), 8.29 (dd, $J = 8.2, 2.1$ Hz, 1H), 7.87 (ddd, $J = 7.9, 7.6, 1.8$ Hz, 1H), 7.39 (ddd, $J = 7.5, 4.8, 1.2$ Hz, 1H) ppm.

5,10,15-Tri([2,2′-bipyridin]-5-yl)-20-(pyridin-4-yl)porphyrinatozinc(II) (**7**)

Pyrrole and 4-pyridinecarboxaldehyde were distilled under reduced pressure before use. A 500 mL round-bottom flask was equipped with a reflux condenser and a magnetic stirring bar. To the flask were added **6** (2.772 g, 15.0 mmol, 2 equiv.) and $EtCO_2H$ (91 mL). The mixture was heated up to 140 °C, and then, to the solution were added 4-pyridinecarboxaldehyde (0.708 mL, 7.52 mmol, 1.00 equiv.) and pyrrole (1.59 mL, 22.9 mmol, 3.05 equiv.). The mixture was stirred at 140 °C for 45 min under aerobic and dark conditions and then cooled to room temperature. H_2O (182 mL) was added, and the mixture was stood for 5 h. The precipitate was collected with suction, washed with H_2O and then CH_3OH, and dried in vacuo to give a purple solid (0.697 g) as a crude mixture of porphyrin derivatives.

A 2 L round-bottom flask was equipped with a magnetic stirring bar. To the flask were added the aforementioned mixture of porphyrin derivatives (0.697 g, 0.798 mmol as porphyrins, 1 equiv.), $CHCl_3$ (520 mL), CH_3OH (52 mL), and $Zn(OAc)_2 \cdot 2H_2O$ (1.493 g, 6.78 mmol, 8.5 equiv. to porphyrins). The mixture was stirred at room temperature for 4.5 h under dark conditions. To the flask was added a 50 mM EDTA·2Na aqueous solution (400 mL, 20 mmol), and the mixture was stirred for 30 min. The organic layer was separated, washed with a 50 mM EDTA·2Na aqueous solution (400 mL) and H_2O (400 mL × 2), dried over Na_2SO_4 and evaporated. The obtained solid was adsorbed onto neutral Al_2O_3 (4 g) by evaporation of a THF solution and purified by neutral Al_2O_3 column chromatography (Merck 40–230 mesh, $\phi = 4$ cm, H = 12 cm, THF/*n*-hexane/pyridine = 100:50:1 to 100:33:2, 100:17:2, 100:0:2). The late fractions were checked by ESI-MS to ensure exclusion of bis(4-pyridyl)porphyrin derivatives. The obtained solid was reprecipitated from $CHCl_3/CH_3OH = 10:1$ adding acetone. The obtained solid was triturated in refluxing $CHCl_3$/acetone = 3:5 for 23 h. The resultant solid was dried in vacuo to afford **7** as a purple solid (215 mg, 0.236 mmol, 4.7%).

^{1}H NMR (DMSO-d_6, 300 K, 500 MHz): δ 9.51 (s, 1H), 9.42 (s, 2H), 9.01 (d, $J = 4.5$ Hz, 2H), 8.98 (d, $J = 4.5$ Hz, 2H), 8.84–8.79 (m, 6H), 8.72–8.67 (m, 6H), 8.60 (br, 2H), 7.99–7.92 (m, 3H), 7.50 (br, 2H), 7.45–7.40 (m, 3H), 6.32 (br, 2H) ppm.

^{13}C NMR (DMSO-*d6*, 300 K, 126 MHz): δ 155.3, 154.5, 152.7, 150.5, 149.7, 149.7, 149.6, 149.6, 148.8, 148.0, 142.0, 138.6, 137.7, 132.3, 132.2, 132.2, 132.1, 129.4, 124.6, 121.0, 118.9, 118.1, 117.1, 117.0 ppm.

mp: > 400 °C.

HR-ESI-MS (positive, $CHCl_3/CH_3OH/HCO_2H$ = 10:1:1, Capillary voltage: 1500 V, Sample cone voltage: 80 V): [**7**·H]$^+$ ($C_{55}H_{34}N_{11}Zn$) *m/z* 912.224 (required, 912.228).

1-Methyl-4-(10,15,20-tri([2,2′-bipyridin]-5-yl)porphyrinatozinc(II)-5-yl)pyridin-1-ium trifluoromethanesulfonate trihydrate (**L1**·OTf·$3H_2O$)

The solid compound **7** (0.215 g, 0.236 mmol, 1 equiv.) was evaporated from its $CHCl_3/CH_3OH$ = 10:1 solution to convert it into a film-like morphology for fast dissolution. A 30 mL pressure tube was equipped with a magnetic stirring bar. To the tube were added the aforementioned film-like solid of **7** (0.215 g, 0.236 mmol, 1 equiv.) dissolved in $CHCl_3/CH_3OH$ = 10:1 (20 mL) and MeI (1.47 mL, 23.6 mmol, 100 equiv.). The tube was sealed and heated at 30 °C for 24 h under dark conditions. The resultant solution was evaporated to dryness. The obtained solid was adsorbed onto neutral Al_2O_3 (1.25 g) by evaporation of a CH_2Cl_2/CH_3OH = 10:1 solution and purified by neutral Al_2O_3 column chromatography (Merck 40–230 mesh, ϕ = 4 cm, H = 12.5 cm, CH_2Cl_2/CH_3OH = 30:1 to 10:1 gradient). The resultant solid was dried in vacuo to afford crude **L1**·I as a green-purple solid (0.162 g, 0.153 mmol, 65%).

A 200 mL round-bottom flask was equipped with a magnetic stirring bar. To the flask were added the aforementioned solid of **L1**·I (0.162 g, 0.153 mmol, 1 equiv.), CH_2Cl_2/CH_3OH = 5:1 (53 mL), and AgOTf (44.5 mg, 0.173 mmol, 1.13 equiv.). The mixture was stirred at room temperature for 18 h under dark conditions. The resultant suspension was evaporated to give a dark solid. The obtained solid was extracted with CH_2Cl_2/CH_3OH = 5:1 (10 mL) by filtration and recrystallized by vapor diffusion of Et_2O. The solid was collected by suction and washed with Et_2O/CH_3OH = 10:1 and then Et_2O. This extraction-recrystallization sequence was repeated once again. The obtained solid was extracted again by centrifugation to strictly exclude any insoluble particles and recrystallized again. The resultant solid was dried in vacuo at 100 °C for 24 h to afford **L1**·OTf·$3H_2O$ as a purple solid (0.154 g, 0.136 mmol, 57% from **7**).

^1H NMR ($CDCl_3/CD_3OD$ = 1:1, 300 K, 500 MHz): δ 9.49 (s, 3H), 9.27 (d, *J* = 6.4 Hz, 2H), 9.08 (d, *J* = 4.7 Hz, 2H), 9.03 (d, *J* = 4.7 Hz, 2H), 9.02 (d, *J* = 4.7 Hz, 2H), 8.96 (d, *J* = 4.7 Hz, 2H), 8.89 (d, *J* = 6.5 Hz, 2H), 8.85 (t, *J* = 4.3 Hz, 3H), 8.80 (d, *J* = 8.0 Hz, 3H), 8.77 (d, *J* = 7.9 Hz, 3H), 8.67 (d, *J* = 7.8 Hz, 3H), 8.11–8.07 (m, 3H), 7.59–7.56 (m, 3H), 4.78 (s, 3H) ppm.

^{13}C NMR ($CDCl_3/CD_3OD$ = 1:1, 300 K, 126 MHz): δ 162.0, 156.3, 156.3, 155.7, 155.7 153.46, 153.43, 151.5, 151.01, 150.84, 149.91, 149.89, 148.8, 143.5, 142.8, 142.8, 139.54, 139.52, 138.4, 138.4, 133.82, 133.63, 133.1, 132.8, 131.1, 125.02,

125.01, 122.6, 122.6, 120.24, 120.21, 119.3, 118.3, 113.3, 48.4 ppm. The resonance from the carbon of TfO^- anion was not observed, probably due to the slow relaxation and splitting by coupling with the ^{19}F nuclei.

^{19}F NMR ($CDCl_3/CD_3OD$ = 1:1, 300 K, 471 MHz): δ −79.2 ppm.

m.p.: > 400 °C.

IR: 574.683 (CF_3), 637.358, 717.39, 750.174, 790.671, 862.989, 993.16, 1029.8, 1066.44, 1157.08 (CF_3), 1206.26, 1222.65 (CF_3), 1254.47, 1339.32, 1367.28, 1433.82, 1457.92, 1525.42, 1546.63, 1571.7, 1587.13, 1636.3 (H_2O), 3053.73 (C-H), 3378.67 (broad, H_2O) cm^{-1}.

HR-ESI-MS (positive, CH_2Cl_2/CH_3OH = 1:5, Capillary voltage: 250 V, Sample cone voltage: 80 V): $[\mathbf{L1}]^+$ ($C_{56}H_{36}N_{11}Zn^+$) *m/z* 926.245 (required, 926.244).

HR-ESI-MS (negative, CH_2Cl_2/CH_3OH = 1:5, Capillary voltage: 100 V, Sample cone voltage: 60 V): $[OTf]^-$ ($CF_3O_3S^-$) *m/z* 148.951 (required, 148.953).

Elemental analysis (calcd., found for $C_{57}H_{42}F_3N_{11}O_6SZn$ [**L**·OTf·3H_2O]): C (60.51, 60.77), H (3.74, 3.72), N (13.62, 13.61).

2.8.2.2 $Co^{II}(OTf)_2 \cdot 6H_2O$

This compound was prepared according to a reported procedure [35]. The product purity was confirmed by elemental analysis.

Elemental analysis (calcd., found for $C_2H_{12}CoF_6O_{12}S_2$ [$Co(OTf)_2 \cdot 6H_2O$]): C (5.16, 5.16), H (2.60, 2.63), N (0.00, 0.00).

2.8.2.3 1-Co^{II}_4

To an NMR tube were added a suspension of **L1**·OTf·3H_2O (0.500 μmol, 1 equiv.) in CD_3CN (100 μL), D_2O (25.0 μL), a solution of $Co(OTf)_2 \cdot 6H_2O$ in CD_3CN (30.0 μL, 0.667 μmol, 22.2 mM, 1.33 equiv.), and CD_3CN (345 μL). The mixture was heated at 70 °C for 4 days.

The yield was estimated to be 65% using *p*-dimethoxybenzene as an internal standard in ^1H NMR.

The ESI-MS spectrum is shown in Fig. 2.15.

2.8.2.4 2-$Co^{III}Co^{II}_3$

To a solution of **1-Co^{II}_4** prepared in the above-mentioned procedure was added a solution of $(NH_4)_2[Ce(NO_3)_6]$ in CD_3CN (10.0 μL, 0.167 μmol, 16.7 mM, 0.333 equiv. to **L1**). The mixture was kept at room temperature for 2 days.

The yield from **1-Co^{II}_4** was estimated to be 83% using *p*-dimethoxybenzene as an internal standard in ^1H NMR. The amount of remaining **1-Co^{II}_4** was also estimated to be 14% in the same manner.

The ESI-MS spectrum is shown in Fig. 2.16.

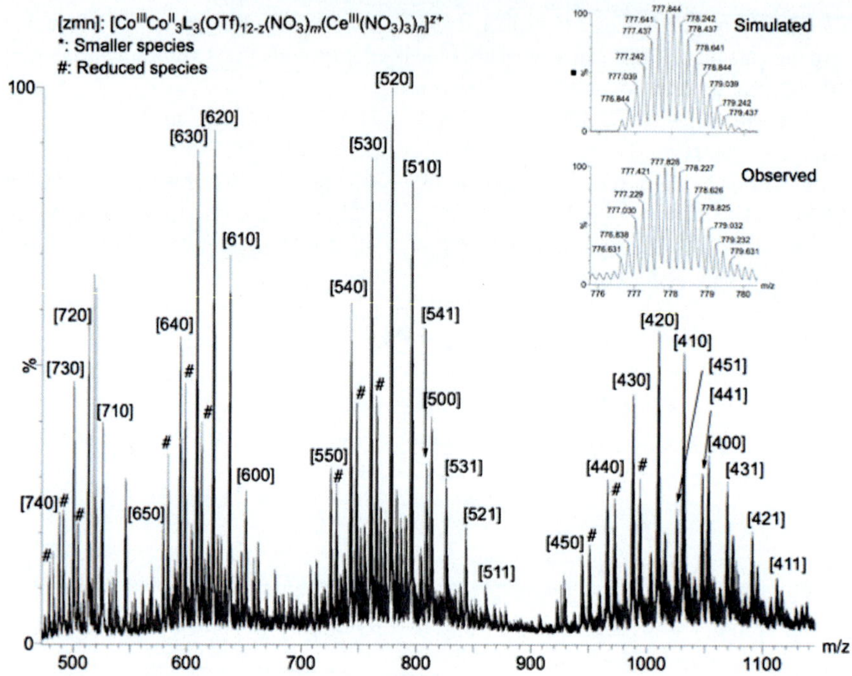

Fig. 2.15 ESI-MS spectrum of the resultant solution containing **1-Co$^{II}_4$** (CD_3CN/D_2O = 19:1, positive, capillary voltage: 1200 V, sample cone voltage: 80 V). The chloride ions are supposed to come from the ESI-MS apparatus. The smaller species may be a product of fragmentation under the measurement condition. Inset shows comparison of simulated and observed isotope patterns for the signal of $[Co^{II}_4\mathbf{L1}_3(OTf)_6]^{5+}$

2.8.2.5 3-Co^{III}Ni$^{II}_3$

To a solution of **2-Co^{III}Co$^{II}_3$** prepared in the above-mentioned procedure was added a solution of $Ni(OTf)_2$ in CD_3CN/D_2O = 9:1 (20.0 µL, 1.00 µmol, 50.0 mM, 2.00 equiv. to **L1**). The mixture was heated at 70 °C for 3 days.

The yield from **1-Co$^{II}_4$** was estimated to be 97% in two steps using *p*-dimethoxybenzene as an internal standard in ^{1}H NMR.

The ESI-MS spectrum is shown in Fig. 2.17.

For XPS measurement and improved synthesis of **4-Co^{II}Ni$^{II}_3$**, the following procedure was added: The resultant solution was evaporated to dryness. The obtained solid was re-dissolved in a mixed solvent of CH_3CN (150 µL) and H_2O (50 µL). To the solution was added an aqueous solution of NH_4PF_6 (250 µL, 25.0 µmol, 100 mM, 50.0 equiv. to **L1**). The formed precipitate was collected and washed with H_2O (800 µL × 2) by centrifugation. The solid was dissolved in dry CH_3CN (500 µL), and all the volatiles were removed under reduced pressure.

The ESI-MS spectrum is shown in Fig. 2.18. The XPS spectrum is shown in Fig. 2.19.

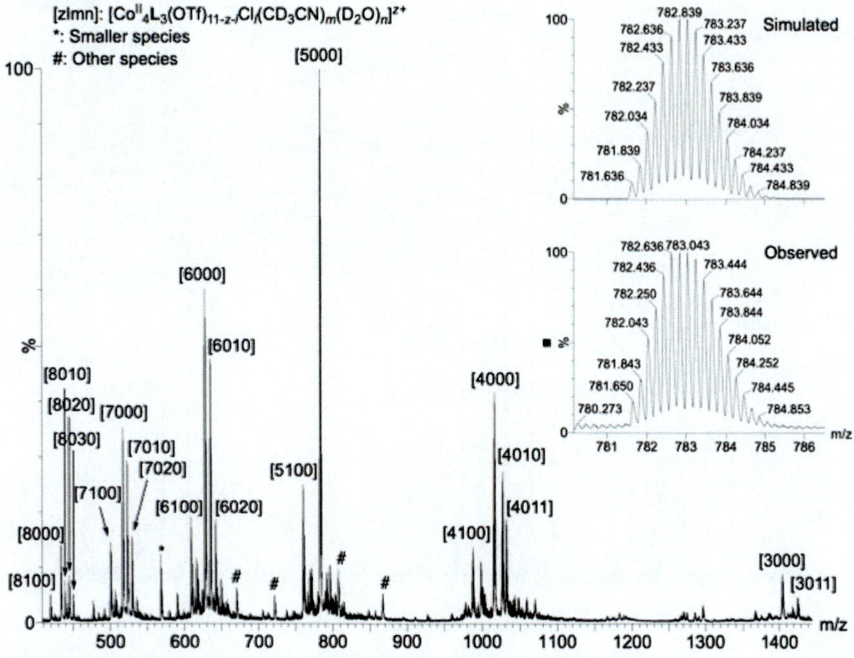

Fig. 2.16 ESI-MS spectrum of the resultant solution containing **2-Co^{III}Co$^{II}_3$** (CD_3CN/D_2O = 19:1, positive, capillary voltage: 3000 V, sample cone voltage: 60 V). The reduced species may be a product of reduction under the measurement condition [44]. Inset shows the comparison of simulated and observed isotope patterns for the signal of [Co^{III}Co$^{II}_3$**L1**$_3$(NO_3)$_2$(OTf)$_5$]$^{5+}$

2.8.2.6 4-Co^{II}Ni$^{II}_3$

Initial attempt: To a solution of **3-Co^{III}Ni$^{II}_3$** prepared in the above-mentioned procedure was added a solution of ^{n}Bu$_4$N·I in CD_3CN (37.5 μL, 0.750 μmol, 20.0 mM, 1.50 equiv. to **L1**). The mixture was heated at 70 °C for 8 h. The yield from **3-Co^{III}Ni$^{II}_3$** was estimated to be 48% using *p*-dimethoxybenzene as an internal standard in ^{1}H NMR. The ESI-MS spectrum is shown in Fig. 2.20.

Improved procedure: The purified sample of **3-Co^{III}Ni$^{II}_3$** prepared in the above-mentioned procedure was dissolved in a mixed solvent of CD_3CN (500 μL) and H_2O (25.0 μL). To the solution was added a solution of ^{n}Bu$_4$N·I in CD_3CN (12.5 μL, 0.625 μmol, 50.0 mM, 0.417 equiv. to the initial amount of **L1**). The reaction mixture was heated at 70 °C for 2 days. The yield from **3-Co^{III}Ni$^{II}_3$** was estimated to be 95% using CH_3OH as an internal standard in ^{1}H NMR. The ESI-MS spectrum is shown in Fig. 2.21.

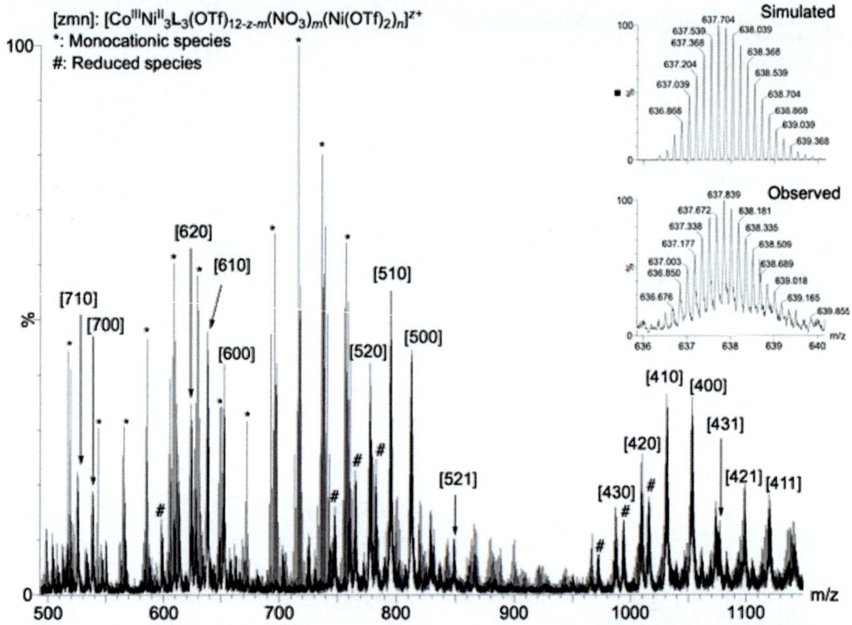

Fig. 2.17 ESI-MS spectrum of the resultant solution containing **3-Co^{III}Ni$^{II}_3$** ($CD_3CN/D_2O = 19:1$, positive, capillary voltage: 3000 V, sample cone voltage: 60 V). The monocationic species are supposed to come from excess Ni^{II} and byproducts Co^{II} and Ce^{III}. The reduced species may be a product of reduction under the measurement condition [44]. Inset shows the comparison of simulated and observed isotope patterns for the signal of $[Co^{III}Ni^{II}_3\mathbf{L1}_3(NO_3)(OTf)_5]^{6+}$

2.8.3 *Miscellaneous Experiments*

2.8.3.1 Stability Test of 4-Co^{II}Ni$^{II}_3$

A solution of **4-Co^{II}Ni$^{II}_3$** prepared in the above-mentioned improved procedure (464 μL, starting from 0.431 μmol of **L1**) was heated at 70 °C for 21 h. DMSO-d_6 (51.6 μL) was added to the solution followed by heating at 70 °C for 14 h.

2.8.3.2 Examination of a One-Step Strategy

To an NMR tube were added a suspension of **L1**·OTf·3H_2O in CD_3CN (100 μL, 0.500 μmol, 1 equiv.), D_2O (25.0 μL), a solution of $Co(OTf)_2{\cdot}6H_2O$ in CD_3CN (5.00 μL, 0.167 μmol, 33.3 mM, 0.333 equiv.), a solution of $Ni(OTf)_2$ in CD_3CN/D_2O = 9:1 (20.0 μL, 0.500 μmol, 25.0 mM, 1.00 equiv.), and CD_3CN (352 μL). The reaction mixture was heated at 70 °C for 5 days.

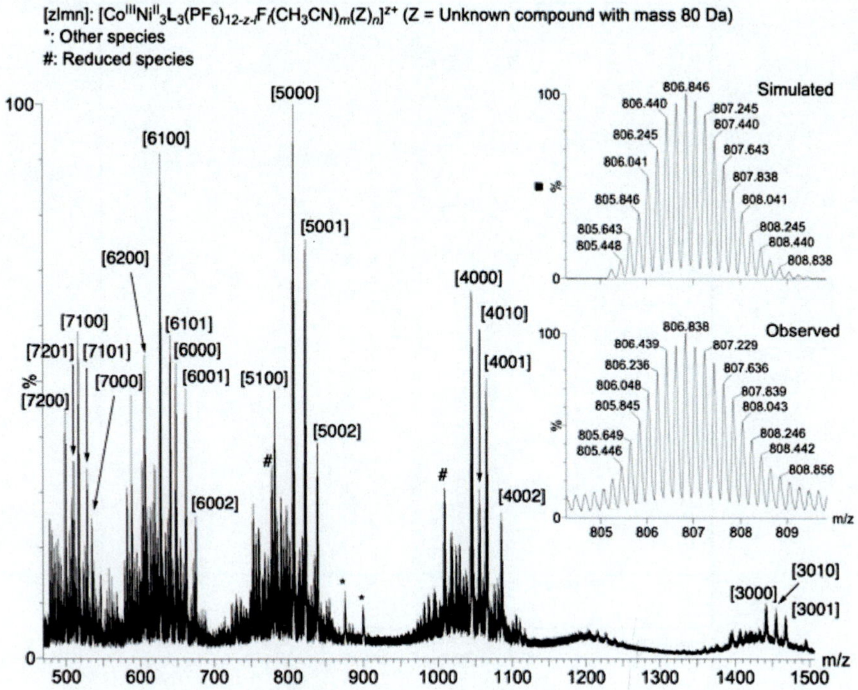

Fig. 2.18 ESI-MS spectrum of the resultant solid containing **3-Co^{III}Ni$^{II}_3$** (CH_3CN, positive, capillary voltage: 3000 V, sample cone voltage: 80 V). The unknown compound Z is supposed to come from the ESI-MS apparatus. The fluoride ions are supposed to be a decomposition product of PF_6^- under the measurement condition. The reduced species may be a product of reduction under the measurement condition [44]. The right inset shows the comparison of simulated and observed isotope patterns for the signal of $[Co^{III}Ni^{II}_3\mathbf{L1}_3(PF_6)_7]^{5+}$

2.8.3.3 Examination of a Transmetalation Strategy Without Redox Switching

To a solution of **1-Co$^{II}_4$** prepared in the above-mentioned procedure was added a solution of $Ni(OTf)_2$ in $CD_3CN/D_2O = 9{:}1$ (40.0 μL, 1.00 μmol, 25.0 mM, 2.00 equiv.). The reaction mixture was heated at 70 °C for 6 days.

2.8.3.4 Examination of a Redox-Switching Strategy Without Transmetalation

To an NMR tube were added a suspension of **L1**·OTf·3H_2O in CD_3CN (100 μL, 0.500 μmol, 1 equiv.), D_2O (25.0 μL), a solution of $Co(OTf)_2{\cdot}6H_2O$ in CD_3CN (5.00 μL, 0.167 μmol, 33.3 mM, 0.333 equiv.), a solution of $Ni(OTf)_2$ in $CD_3CN/D_2O = 9{:}1$ (20.0 μL, 0.500 μmol, 25.0 mM, 1.00 equiv.), and CD_3CN (352 μL). The mixture was heated at 70 °C for 5 days. To this solution was added a solution of

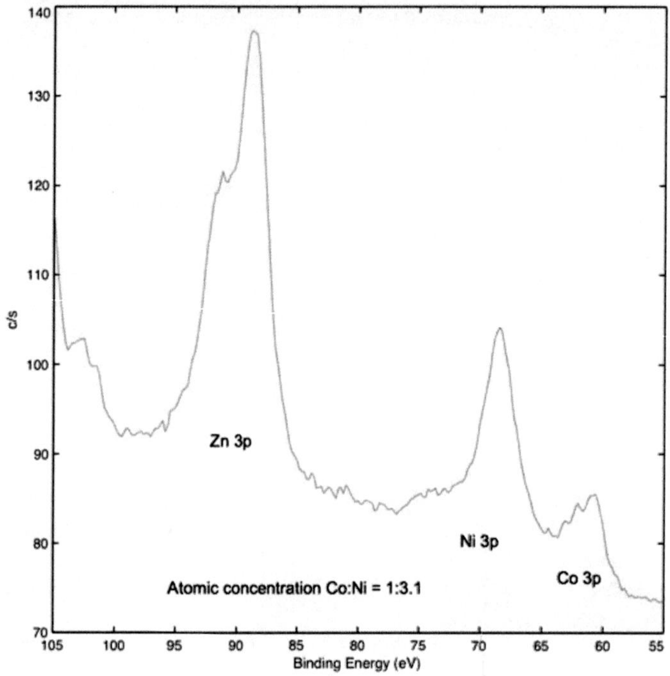

Fig. 2.19 Core-level XPS spectrum of the resultant solid containing **3-$Co^{III}Ni^{II}_3$**

$(NH_4)_2[Ce(NO_3)_6]$ in CD_3CN (23.0 μL, 0.167 μmol, 7.26 mM, 0.333 equiv.). The reaction mixture was heated at 70 °C for 4 days.

2.8.4 Single-Crystal XRD Analyses

2.8.4.1 1-Co^{II}_4

A single crystal of **1-Co^{II}_4** was not obtained from the sample mentioned above. Crystals suitable for X-ray analysis were obtained as described below.

To an NMR tube were added a solution of **L1**·OTf·$3H_2O$ in $CDCl_3/CD_3OD/D_2O$ = 10:10:1 (100 μL, 0.500 μmol, 5.00 mM, 1 equiv.), a solution containing $Co(NO_3)_2·6H_2O$ in the same mixed solvent (30.0 μL, 0.690 μmol, 23.0 mM, 1.38 equiv.), and the same mixed solvent (370 μL). The reaction mixture was heated at 50 °C for 2 days. The solution contained **1-Co^{II}_4** as the main product as confirmed by the ^1H NMR and ESI-MS measurements. To the mixture was added a solution of $La(OTf)_3·xH_2O$ in CH_3OH (10.0 μL, 0.250 μmol, 25.0 mM, 0.500 equiv.) to facilitate crystallization. Slow vapor diffusion of Et_2O to this solution afforded some single crystals suitable for X-ray analysis which were found in the powdery crude product.

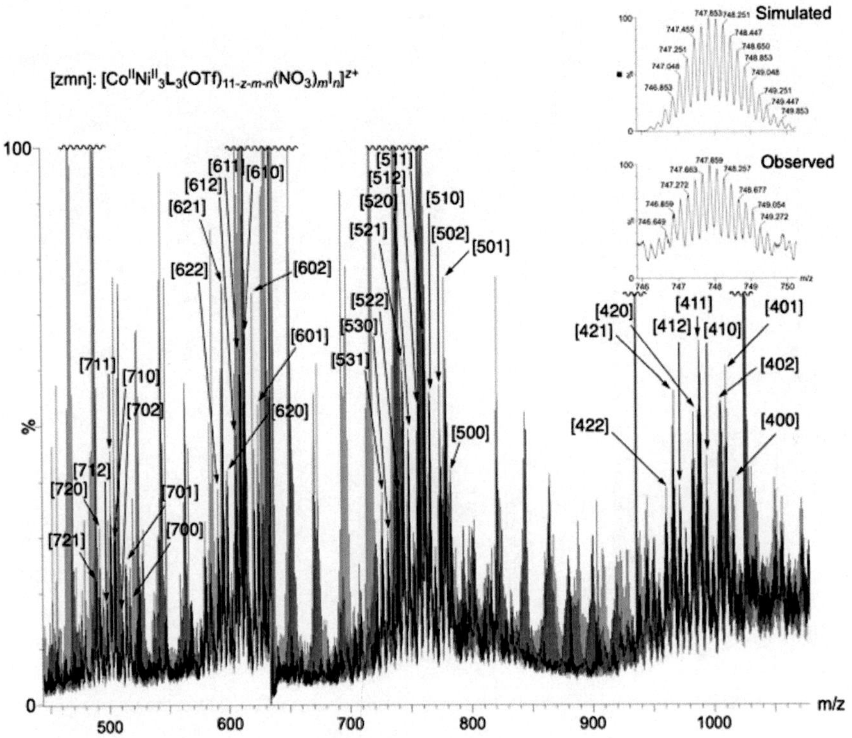

Fig. 2.20 ESI-MS spectrum of the resultant solution containing **4-Co^{II}Ni$^{II}_{3}$** (CD_3CN/D_2O = 19:1, positive, capillary voltage: 3000 V, sample cone voltage: 80 V). Numerous monocationic species were observed, presumably because of the excess Ni^{II} and byproducts $^{n}Bu_4N^+$, Co^{II}, and Ce^{III}. The right inset shows the comparison of simulated and observed isotope patterns for the signal of $[Co^{II}Ni^{II}_{3}\mathbf{L1}_3(NO_3)_2(OTf)_4]^{5+}$

Crystal data for $[Co_4\mathbf{L1}_3(OH_2)_6(HOCH_3)_3][La(NO_3)_6]_{0.5}(OTf)_{9.5}$: $C_{169.96}H_{113.88}Co_4La_{0.5}N_{36}O_{16.96}Zn_3$, Fw = 3433.04, violet, plate, 0.099 × 0.366 × 0.400 mm^3, trigonal, space group P-3c1 (#165), a = 21.3538(4) Å, b = 21.3538(4) Å, c = 58.2741(18) Å, $\alpha = 90°$, $\beta = 90°$, $\gamma = 120°$, V = 23012.1(11) Å^3, Z = 4, ρ_{calcd} = 0.991 g·cm^{-3}, T = 93 K, λ(CuKα) = 1.54184 Å, $2\theta_{max}$ = 136.502°, 65158/14055 reflections collected/unique (R_{int} = 0.0395), R_1 = 0.1058 ($I > 2\sigma(I)$), wR_2 = 0.3319 (for all data), GOF = 1.372, largest diff. peak and hole 1.23/-0.54 e·Å^{-3}. CCDC deposit number 1896478.

At one of the pyridine rings, thermal parameter restraints (RIGU) were applied to facilitate anisotropic refinement. The zinc atom in the porphyrin ring was modeled as disordered over inside and outside positions. The axial CH_3OH molecule bound to the zinc center was modeled in the same occupancy as the zinc atom, although the axial ligand of the minor disorder component could not be located on a d-Fourier

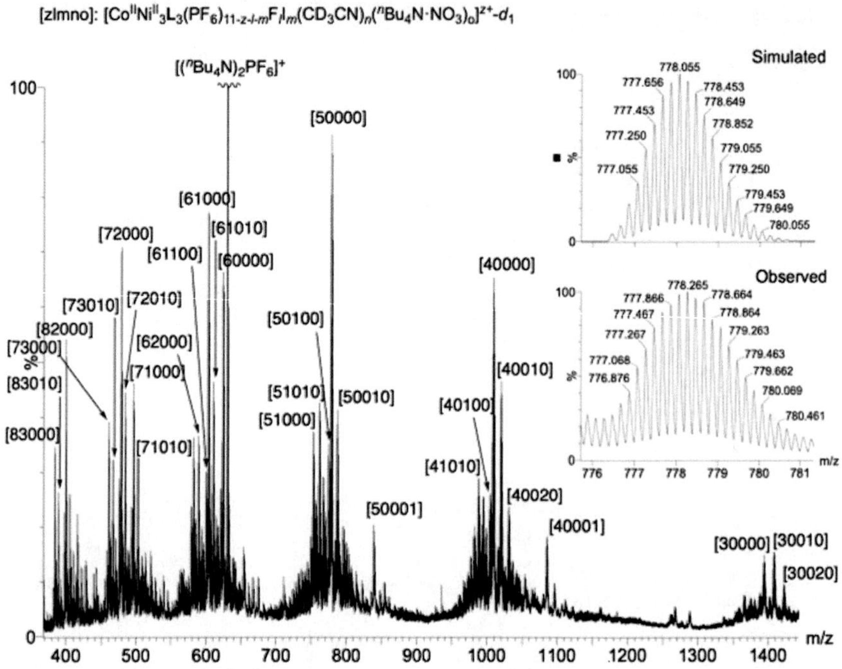

Fig. 2.21 ESI-MS spectrum of the resultant solution containing **4-Co^{II}Ni$^{II}_3$** in the improved procedure (CD_3CN/D_2O = 20:1, positive, capillary voltage: 3000 V, sample cone voltage: 80 V). The right inset shows the comparison of simulated and observed isotope patterns for the signal of $[Co^{II}Ni^{II}_3$**L**$_3(PF_6)_6]^{5+}$-d_1. The deuteration of the ligand was supposed to occur at the pyridinium *N*-methyl position under the reaction or measurement condition

map. The NO_3^- ligand and the nearby *N*-methylpyridinium and porphyrin moieties were modeled as disordered over two positions. Due to the close positions of the disordered components, the bond lengths and angles were restrained to be similar to each other (SAME), and thermal parameter restraints (SIMU, RIGU) were applied to facilitate anisotropic refinement. Although 11 anions should be contained as shown in the chemical formula, only 0.5 $[La(NO_3)_6]^{3-}$ ion was found. The remaining anions were supposed to be so heavily disordered that they could not be located on a *d*-Fourier map. After successive trials, solvent accessible voids of 9544.7 Å^3 in total (calculated by solvent mask) were left unfilled, in which anions and solvents ($CHCl_3$, CH_3OH, H_2O, and Et_2O) were supposed to be heavily disordered. At the last stage of refinement, the reflection data modified by solvent mask were used. The solvent accessible void contained 3215 electrons per unit cell. This equates to 10.9 TfO$^-$ (each with 74 electrons) for each molecule of **1-Co$^{II}_4$**, where $Z = 4$. This diffuse electron density exceeds well the 9.5 TfO$^-$ anions necessary to satisfy the 11 + charge of **1-Co$^{II}_4$**, leaving some room for the contribution of solvent molecules.

2.8.4.2 2-$Co^{III}Co^{II}_3$

A single crystal of **2-$Co^{III}Co^{II}_3$** was not obtained from the sample mentioned above. Crystals suitable for X-ray analysis was obtained as described below.

To an NMR tube were added a solution of **L1**·OTf·3H_2O in $CDCl_3/CD_3OD/D_2O$ = 10:10:1 (100 μL, 0.500 μmol, 5.00 mM, 1 equiv.), a solution of $Co(NO_3)_2$·6H_2O in the same mixed solvent (30.0 μL, 0.690 μmol, 23.0 mM, 1.38 equiv.), and the same mixed solvent (370 μL). The reaction mixture was heated at 50 °C for 3 days to obtain a dark green solution. To the mixture was added a solution of $(NH_4)_2[Ce(NO_3)_6]$ in CD_3OD (10.0 μL, 0.167 μmol, 25.0 mM, 0.333 equiv.). The mixture was kept at room temperature for 5 h. The solution was found to contain the expected complex as the main product as confirmed by the ^1H NMR and ESI-MS measurements. Slow vapor diffusion of Et_2O to this solution afforded some single crystals suitable for X-ray analysis which were found in the powdery crude product.

Crystal data for $[Co_4\mathbf{L1}_3(OH_2)_6(HOCH_3)_3][Ce(NO_3)_6]_{0.5}(OTf)_{10.5}$: $C_{170.68}H_{116.03}Ce_{0.5}Co_4N_{36}O_{17.67}Zn_3$, $Fw = 3455.80$, violet, plate, 0.067 × 0.188 × 0.420 mm^3, trigonal, space group P-3c1 (#165), $a = 21.2460(4)$ Å, $b = 21.2460(4)$ Å, $c = 58.2102(12)$ Å, $\alpha = 90°$, $\beta = 90°$, $\gamma = 120°$, $V = 22755.4(10)$ Å^3, $Z = 4$, $\rho_{calcd} = 1.009$ g·cm^{-3}, $T = 93$ K, λ(CuKα) = 1.54184 Å, $2\theta_{max} = 136.376°$, 54339/13847 reflections collected/unique ($R_{int} = 0.0491$), $R_1 = 0.0994$ ($I > 2\sigma(I)$), $wR_2 = 0.3230$ (for all data), GOF = 1.147, largest diff. peak and hole 0.99/-0.89 e·Å^{-3}. CCDC deposit number 1896498.

At one of the bipyridine moieties, thermal parameter restraints (RIGU) were applied to facilitate anisotropic refinement. The zinc atom in the porphyrin ring was modeled as disordered over inside and outside positions. The axial CH_3OH molecule bound to the zinc center was modeled in the same occupancy as the zinc atom, although the axial ligand of the minor disorder component could not be located on a *d*-Fourier map. The NO_3^- ligand and the nearby *N*-methylpyridinium and porphyrin moieties were modeled as disordered over two positions. Due to the close positions of the disordered components, the bond lengths and angles were restrained to be similar to each other (SADI, SAME), and thermal parameter restraints (SIMU, RIGU) were applied to facilitate anisotropic refinement. Although 12 anions should be contained as shown in the chemical formula, only 0.5 $[Ce(NO_3)_6]^{3-}$ ion was found. The remaining anions were supposed to be so heavily disordered that they could not be located on a *d*-Fourier map. After successive trials, solvent accessible voids of 9360.8 Å^3 in total (calculated by solvent mask) were left unfilled, in which anions and solvents ($CHCl_3$, CH_3OH, H_2O, and Et_2O) were supposed to be heavily disordered. At the last stage of refinement, the reflection data modified by solvent mask were used. The solvent accessible void contained 3352.4 electrons per unit cell. This equates to 11.3 TfO$^-$ (each with 74 electrons) for each molecule of **2-$Co^{III}Co^{II}_3$**, where $Z = 4$. This diffuse electron density exceeds well the 10.5 TfO$^-$ anions necessary to satisfy the 12 + charge of **2-$Co^{III}Co^{II}_3$**, leaving some room for the contribution of solvent molecules.

2.8.4.3 3-$Co^{III}Ni^{II}_3$

A single crystal of **3-$Co^{III}Ni^{II}_3$** was not obtained from the sample mentioned above. Crystals suitable for X-ray analysis were obtained as described below.

To a glass vial was added **L1**·OTf·3H_2O (23.2 mg, 20.5 μmol, 1 equiv.), CH_2Cl_2/CH_3OH = 5:1 (5 mL), and $Bu_4N \cdot NO_3$ (65 mg, 0.21 mmol, 10 equiv.). The solution was filtered and washed with CH_2Cl_2/CH_3OH = 5:1. Vapor of Et_2O was slowly diffused to the combined filtrate. The solid was collected by filtration, washed with Et_2O/CH_2Cl_2 = 5:1 and then CH_2Cl_2, and dried in vacuo to give crude **L1**·NO_3 as a dark violet solid (20.8 mg).

To an NMR tube were added a solution of **L1**·NO_3 in $CDCl_3/CD_3OD/D_2O$ = 10:10:1 (250 μL, 0.500 μmol, 2.00 mM, 1 equiv.), a solution of $Co(NO_3)_2 \cdot 6H_2O$ in the same mixed solvent (18.9 μL, 0.667 μmol, 35.3 mM, 1.33 equiv.), and the same mixed solvent (231 μL). The mixture was heated at 50 °C for 27 h to obtain a dark green solution. To the mixture were added a solution of $(NH_4)_2Ce(NO_3)_6$ solution in the same mixed solvent (10.0 μL, 0.167 μmol, 16.7 mM, 0.333 equiv.) and a solution of $Ni(OTf)_2$ in CD_3OD/D_2O = 1:1 (10.0 μL, 0.500 μmol, 50.0 mM, 1.00 equiv.). The reaction mixture was heated at 50 °C for 3 days. To the solution was added one more aliquot of the same $Ni(OTf)_2$ solution (10.0 μL, 0.500 μmol, 1.00 equiv.). The reaction mixture was heated at 50 °C for 2 days. The solution contained **3-$Co^{II}Ni^{II}_3$** as the main product as confirmed by ^{1}H NMR spectroscopy and ESI-MS. To the aliquot of the solution (432 μL) was added DMSO-d_6 (100 μL). Slow vapor diffusion of Et_2O to this solution afforded some single crystals suitable for X-ray analysis which were found in the powdery crude product.

Crystal data for [$CoNi_3$**L1**$_3Cl_3(dmso)_6](OTf)_9 \cdot$DMSO: $C_{179.77}H_{140.32}Cl_3CoF_3N_{33}Ni_3O_{9.7}S_{6.39}Zn_3$, Fw = 3717.33, violet, plate, 0.1 × 0.1 × 0.02 mm^3, trigonal, space group R-3 (#148), a = 21.5526(5) Å, b = 21.5526(5) Å, c = 83.834(3) Å, $\alpha = 90°$, $\beta = 90°$, $\gamma = 120°$, V = 33725(2) Å^3, Z = 6, ρ_{calcd} = 1.098 g·cm^{-3}, T = 93 K, λ(MoKα) = 0.71073 Å, $2\theta_{max}$ = 57.67°, 46042/16339 reflections collected/unique (R_{int} = 0.0485), R_1 = 0.0595 ($I > 2\sigma(I)$), wR_2 = 0.1797 (for all data), GOF = 1.051, largest diff. peak and hole 0.80/−0.71 e·Å^{-3}. CCDC deposit number 1898214.

As Co and Ni atoms have similar electron densities, assignment of these atoms was conducted by consideration of the M–N bond distances as discussed in the main text.

At the methyl group of the *N*-methylpyridinium moiety, thermal parameter restraints (RIGU) were applied to facilitate anisotropic refinement. The zinc atom in the porphyrin ring was modeled as disordered over inside and outside positions. Due to the close positions and the highly biased occupancy of the disordered atoms, the bond lengths and angles were restrained to be similar to typical values (DFIX, DANG), and a thermal parameter constraint (EADP) was applied to facilitate anisotropic refinement. The axial DMSO bound to the zinc center was modeled in the same occupancy as the zinc atom with disorder over two positions. Due to the close positions of the disordered components, thermal parameter restraints (SIMU, RIGU) were applied to facilitate anisotropic refinement. The axial ligand of the minor disorder component of zinc could not be located on a *d*-Fourier map. One of the ligands coordinating to the nickel center was modeled as disordered over H_2O and DMSO molecules. Due to the diffuse electron densities of the disordered components, the bond lengths and angles were restrained to be symmetric (SAME), and thermal parameter restraints (RIGU) were applied to facilitate anisotropic refinement. The DMSO molecule located inside the complex was modeled as disordered around the 3-fold axis. Due to the severe overlapping of electron density around the 3-fold axis, the bond lengths were restrained to be similar to typical values (DFIX) and the molecular geometry to be symmetric (SAME), and thermal parameter restraints (SIMU, RIGU) were applied to facilitate refinement. Although nine anions should be contained as shown in the chemical formula, only one TfO^- ion was found. For this ion, thermal parameter restraints (RIGU) were applied to facilitate anisotropic refinement. The remaining anions were supposed to be so heavily disordered that they could not be located on a *d*-Fourier map. After successive trials, solvent accessible voids of 11136 $Å^3$ in total (calculated by SQUEEZE) were left unfilled, in which anions and solvents ($CHCl_3$, CH_3OH, H_2O, DMSO, and Et_2O) were supposed to be heavily disordered. At the last stage of refinement, the reflection data modified by SQUEEZE were used. The solvent accessible void contained 4196 electrons per unit cell. This equates to 9.45 TfO^- (each with 74 electrons) for each molecule of **3-$Co^{III}Ni^{II}_3$**, where $Z = 6$. This diffuse electron density exceeds well the 8 TfO^- anions necessary to satisfy the 12+ charge of **3-$Co^{III}Ni^{II}_3$**, leaving some room for the contribution of solvent molecules.

References

1. Balzani V, Juris A, Venturi M (1996) Chem Rev 96:759–833
2. Iki N (2011) Supramol Chem 23:160–168
3. Ostrowska M, Fritsky IO, Gumienna-Kontecka E, Paclishchuk AV (2016) Coord Chem Rev 327–328:304–322
4. Aboshyan-Sorgho L, Besnard C, Pattison P, Kittilstved KR, Aebischer A, Bunzli JC, Hauser A, Piguet C (2011) Angew Chem Int Ed 50:4108–4112
5. Huang Y-G, Jiang F-L, Hong M-C (2008) Coord Chem Rev 253:2814–2834
6. Shimada T, Okazawa A, Kojima N, Yoshii S, Nojiri H, Ishida T (2011) Inorg Chem 50:10555–10557
7. Buchwalter P, Jacky R, Braunstein P (2015) Chem Rev 115:28–126
8. Sasai H, Suzuki T, Itoh N, Tanaka K, Date T, Okamura K, Shibasaki M (1993) J Am Chem Soc 115:10372–10373
9. Que L Jr, Tolman WB (eds) (2003) Comprehensive coordination chemistry II, vol 8. Elsevier, Amsterdam
10. Strange RW, Antonyuk SV, Hough MA, Doucette PA, Valentine JS, Hasnain SS (2006) J Mol Biol 356:1152–1162
11. Zhang Y-Y, Gao W-X, Lin L, Jin G-X (2016) Coord Chem Rev 344:323–344
12. Li H, Yao Z-J, Liu D, Jin G-X (2015) Coord Chem Rev 293–294:139–157
13. Sun X, Johnson DW, Caulder DL, Powers RE, Raymond KN, Wong EH (1999) Angew Chem Int Ed 38:1303–1307
14. Piguet C, Bernardinelli G, Williams AF, Bocquet B (1995) Angew Chem Int Ed Engl 34:582–584

15. Hiraoka S, Tanaka T, Shionoya M (2006) J Am Chem Soc 128:13038–13039
16. Kumar G, Gupta R (2013) Chem Soc Rev 42:9403–9453
17. Li L, Fanna DJ, Shepherd ND, Lindoy LF, Li F (2015) J Incl Phenom Macrocycl Chem 82:3–12
18. Lintvedt RL, Ahmad N (1982) Inorg Chem 21:2356–2359
19. Davies G, El-Sayed MA, El-Toukhy A (1992) Chem Soc Rev 21:101–104
20. Carnes ME, Collins MS, Johnson DW (2014) Chem Soc Rev 43:1825–1834
21. Cai GZ, Davies G, El-Sayed MA, El-Toukhy A, Gilbert TR, Onan KD (1986) Inorg Chem 25:1935–1940
22. Mensinger ZL, Zakharov LN, Johnson DW (2008) Acta Cryst. E64:i8–i9
23. Simler T, Braunstein P, Danopoulos AA (2015) Angew Chem Int Ed 54:13691–13695
24. Ai P, Monakhov KY, van Leusen J, Kögerler P, Gourlaouen C, Tromp M, Welter R, Danopoulos AA, Braunstein P (2018) Chem Eur J 24:8787–8796
25. Kamunde-Devonish MK, Jackson MN Jr, Mensinger ZL, Zakharov LN, Johnson DW (2014) Inorg Chem 53:7101–7105
26. Givaja G, Castillo O, Mateo E, Gallego A, Gómez-García CJ, Calzolari A, di Felice R, Zamora F (2012) Chem Eur J 18:15476–15484
27. Pittala S, Kittilstved KR (2015) Cation exchange in small ZnS and CdS molecular analogues. Inorg Chem 54:5757–5767
28. Echegoyen L, Perez-Cordero E (1994) Redox chemistry of metal ion complexes: preparation of new materials. In: Transition metals in supramolecular chemistry. Springer, Dordrecht
29. Burke MJ, Nichol GS, Lusby PJ (2016) J Am Chem Soc 138:9308–9315
30. Endo K, Ube H, Shionoya M (2020) J Am Chem Soc 142:407–416
31. Brisig B, Constable EC, Housecroft CE (2007) New J Chem 31:1437–1447
32. Huang TLJ, Brewer DG (1981) Can J Chem 59:1689–1700
33. Hall BR, Manck LE, Tidmarsh IS, Stephenson A, Taylor BF, Blaikie EJ, Vander Griend DA, Ward M (2011) d. Dalton Trans 40:12132–12145
34. Turega S, Whitehead M, Hall BR, Haddow MF, Hunter CA, Ward MD (2012) Chem Commun 48:2752–2754
35. Riddell IA, Smulders MM, Clegg JK, Hristova YR, Breiner B, Thoburn JD, Nitschke JR (2012) Nat Chem 4:751–756
36. Haynes WM (2013) CRC handbook of chemistry and physics, 94th edn. CRC Press, Boca Raton
37. Sasaki Y, Kato H, Kudo A (2013) J Am Chem Soc 135:5441–5449
38. Groom CR, Bruno IJ, Lightfoot MP, Ward SC (2016) Acta Cryst B72:171–179
39. Smith RM, Martell AE (1976) Critical stability constants. Plenum Press, New York
40. Nakamura T, Kaneko Y, Nishibori E, Nabeshima T (2017) Nat Commun 8:129
41. Dolomanov OV, Bourhis LJ, Gildea RJ, Howard JAK, Puschmann HJ (2009) Appl Cryst 42:339–341
42. Sheldrick GM (2015) Acta Cryst C 71:3–8
43. Nakamura T, Ube H, Shiro M, Shionoya M (2013) Angew Chem Int Ed 52:720–723
44. Gianelli L, Amendola V, Fabbrizzi L, Pallavicini P, Mellerio GG (2001) Rapid Commun Mass Spectrom 15:2347–2353

Chapter 3
Tetrahedral Chiral-at-Metal Zn^{II} Complex

Abstract A metal center of a metal complex can be a chirality center in some cases, which are called chiral-at-metal complexes. Chiral-at-metal complexes are attractive because it can act as enantioselective catalyst, chiroptical material, and so on in the absence of chiral ligands. However, there are two problems in chiral-at-metal complexes without chiral ligands. Firstly, pure enantiomers are usually obtained only by separation, and enantioselective synthesis is rarely reported. Secondly, such complexes are mostly limited to octahedral, relatively inert metal complexes, while tetrahedral complexes are supposed to racemize quickly in the absence of chiral ligands. Therefore, it remains as a challenging task to enantioselectively synthesize a stable tetrahedral chiral-at-metal complex. Since enantiomers have the same thermodynamic property, enantioselective synthesis requires kinetic control. Here, I report enantioselective synthesis of a stable tetrahedral chiral-at-metal Zn^{II} complex by three-step sequence: racemic complexation, asymmetric induction, and replacement of a chiral ligand. It was also necessary to design a special ligand to stabilize the complex against racemization. In addition, the practical utility of the complex was demonstrated in application to enantioselective catalysis.

Keywords Chiral-at-metal complex · Enantioselective synthesis · Chiral auxiliary · Racemization · Enantioselective catalysis

3.1 Introduction

Chirality, the property of the objects which are not superimposable on their mirror images, is important in any fields of chemistry. If a molecule is chiral, its structure and properties need to be considered with regard to its two enantiomers. The differences between the enantiomers are important in various aspects: biological activities such as pharmaceutical [1], agrochemical [2], or fragrant [3] effects; chiroptical properties such as circularly polarized luminescence [4, 5]; and multicomponent systems such as self-assembly [6] and polymer tacticity [7] (Fig. 3.1). Apart from these artificial chemicals, biological systems also utilize discrimination of enantiomers as the key to the construction of well-regulated chemical networks and functional constituents

K. Endo, *Kinetically Controlled Stepwise Syntheses of a Heterometallic Complex and a Tetrahedral Chiral-at-Metal Complex*, Springer Theses,
https://doi.org/10.1007/978-981-16-1163-6_3

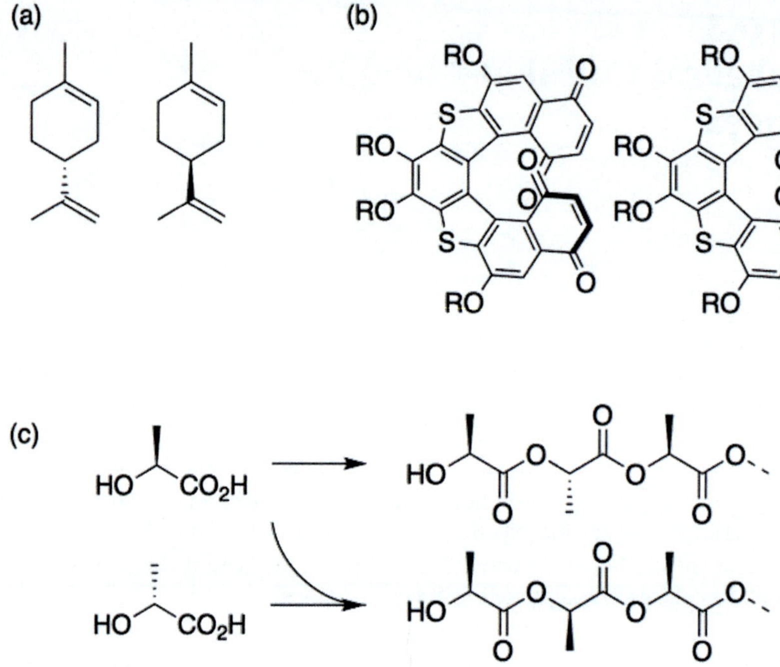

Fig. 3.1 Examples of the chiral molecules whose differences between the enantiomers are important. **a** Bioactive compounds. **b** Compounds active for circularly polarized luminescence [4]. **c** Monomers for polymers with different tacticities

[8]. For example, incorporation of just one molecule of the wrong enantiomer of an amino acid can ruin the function of a protein [9].

In both artificial and natural chiral molecular entities, the source of chirality is usually a group of nonmetal atoms, exemplified by a tetrahedral carbon atom with four different substituents (Fig. 3.2a–c). On the other hand, though less common, a metal center in a metal complex can also be a chirality source, as was first discovered

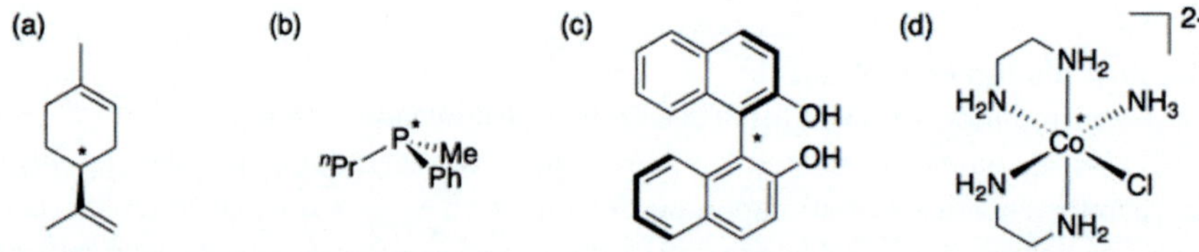

Fig. 3.2 Examples of sources of chirality: **a** tetrahedral carbon atom with four different substituents; **b** trigonal-pyramidal phosphine atom with three different substituents [11]; **c** carbon–carbon bond whose rotation is restricted; **d** octahedral metal center with two *cis*-bidentate ligands and two monodentate ligands in a *cis* arrangement [10]. The chirality sources are labeled with asterisks

in 1911 by Werner (Fig. 3.2d) [10]. Recently, such a complex is called a "chiral-at-metal" complex and getting a lot more attention. Since a metal center can exhibit various chemical and physical functions, chiral-at-metal complexes bearing metal centers as integrated function-and-chirality sources are an attractive structural motif for designing a chiral molecular entity.

After Werner's discovery, various kinds of chiral-at-metal complexes have been reported. Their coordination geometry ranges from octahedral [10] to tetrahedral [12], square-planar [13], trigonal-bipyramidal [14], half-sandwich (pseudooctahedral) [15], tricapped-trigonal-prismatic [16], and capped-square-antiprismatic [17] (Fig. 3.3).

To utilize a chiral-at-metal complex, the absolute configuration of its metal center should be fixed. In previous researches, the absolute configurations of metal centers were usually controlled in the presence of other chiral components (ligands [19–23] or counterions [24–27]) under thermodynamic equilibrium (Fig. 3.4a). On the other hand, some metal centers can be configurationally stable and obtained in enantiopure forms even in the absence of other chiral components [10, 28] (Fig. 3.4b). Such configurationally stable chiral metal centers are attractive because the whole system can be simpler than those with other chiral components.

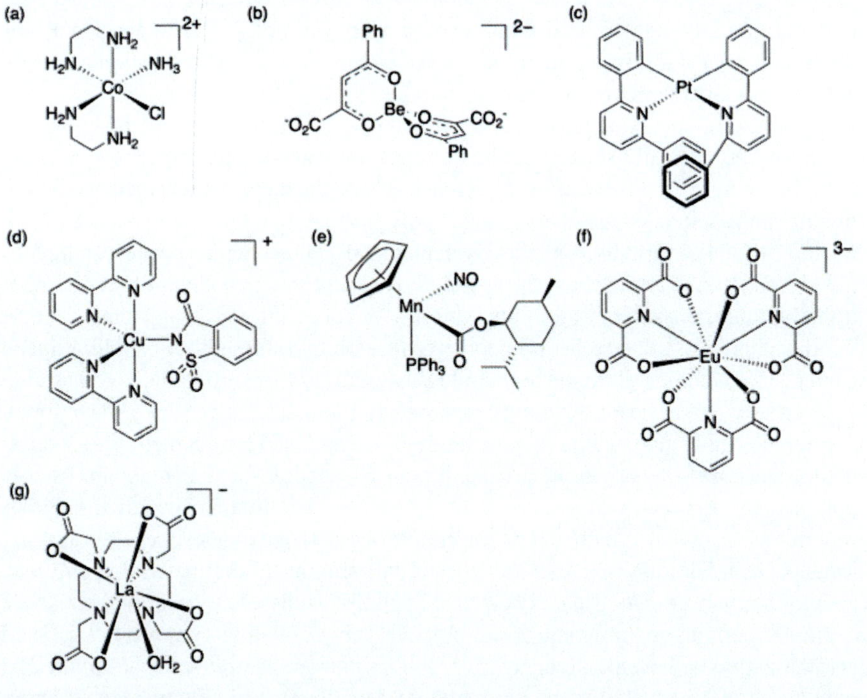

Fig. 3.3 Examples of chiral-at-metal complexes with various coordination geometries: **a** octahedral [10]; **b** tetrahedral [13]; **c** square-planar [14]; **d** trigonal-bipyramidal [15]; **e** half-sandwich (pseudooctahedral) [18]; **f** tricapped-trigonal-prsimatic [17]; and **g** capped-square-antiprismatic [19]

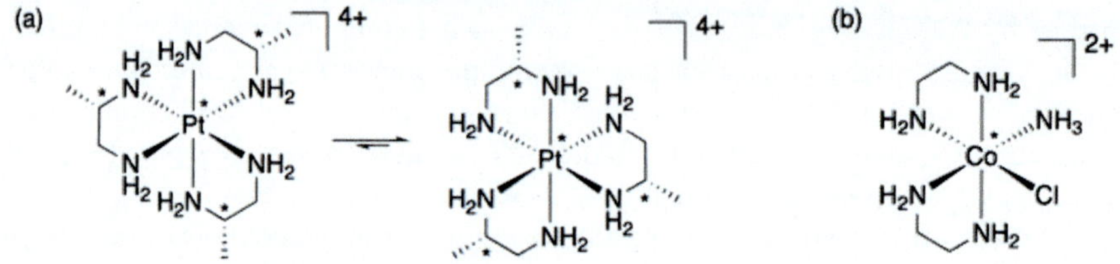

Fig. 3.4 Examples of fixation of the absolute configurations of the metal centers: **a** controlling the absolute configuration of a metal center with chiral ligands under thermodynamic control [29]; **b** a metal center which is configurationally stable without any other chiral compounds [10]. The chirality centers are labeled with asterisks

The usefulness of configurationally stable chiral metal centers has been shown in applications to a variety of fields including enantioselective catalysis [30–32], circularly polarized luminescence [33], chiral magnetic materials [34], enantioseparation [35], and biological purposes [35] (Fig. 3.5).

However, the previous researches on configurationally stable chiral metal centers have been focused on octahedral or half-sandwich geometry with d^3 or d^6 metals. The researches on tetrahedral geometry are significantly retarded, although it is directly related to a carbon center. Specifically, the use of tetrahedral geometry for configurationally stable chiral metal centers is confronted by three problems: low configurational stability, a limited number of enantioselective construction methods, and the lack of researches on catalytic activity.

Firstly, while many octahedral and half-sandwich chiral metal centers are known to be configurationally stable, tetrahedral ones are usually not. One reason is faster ligand exchange because of more open space around the metal centers than octahedral or half-sandwich ones. Another reason is polytopal isomerization into square-planar geometry [37]. Currently, only low-valent metal centers with π-acceptor ligands [38, 39, 41] and polynuclear systems [42–50] are reported to be relatively configurationally stable, and still their stability is modest or not confirmed at high temperatures. To make better use of tetrahedral geometry for configurationally stable chiral metal centers, a new approach to make them highly configurationally stable is required.

Secondly, while various methods are reported for enantioselective construction of configurationally stable chiral octahedral metal centers [51], such methods for tetrahedral ones are severely limited. Enantiopure tetrahedral metal centers are usually obtained by separation of a racemic or diastereomeric mixture, which sometimes requires laborious search for separation conditions [13] or expensive separation apparatus [47]. In addition, this method loses a half amount of the material if only one enantiomer is necessary. Enantioselective synthesis is therefore desirable, but so far available methods are limited to only chiral guest encapsulation–removal [52, 53] and crystallization with a chiral reagent [54]. To facilitate preparation of enantiopure and configurationally stable tetrahedral metal centers, its enantioselective construction methods should be further developed.

Thirdly, while configurationally stable chiral octahedral metal centers have been extensively utilized for enantioselective catalytic conversion of organic compounds

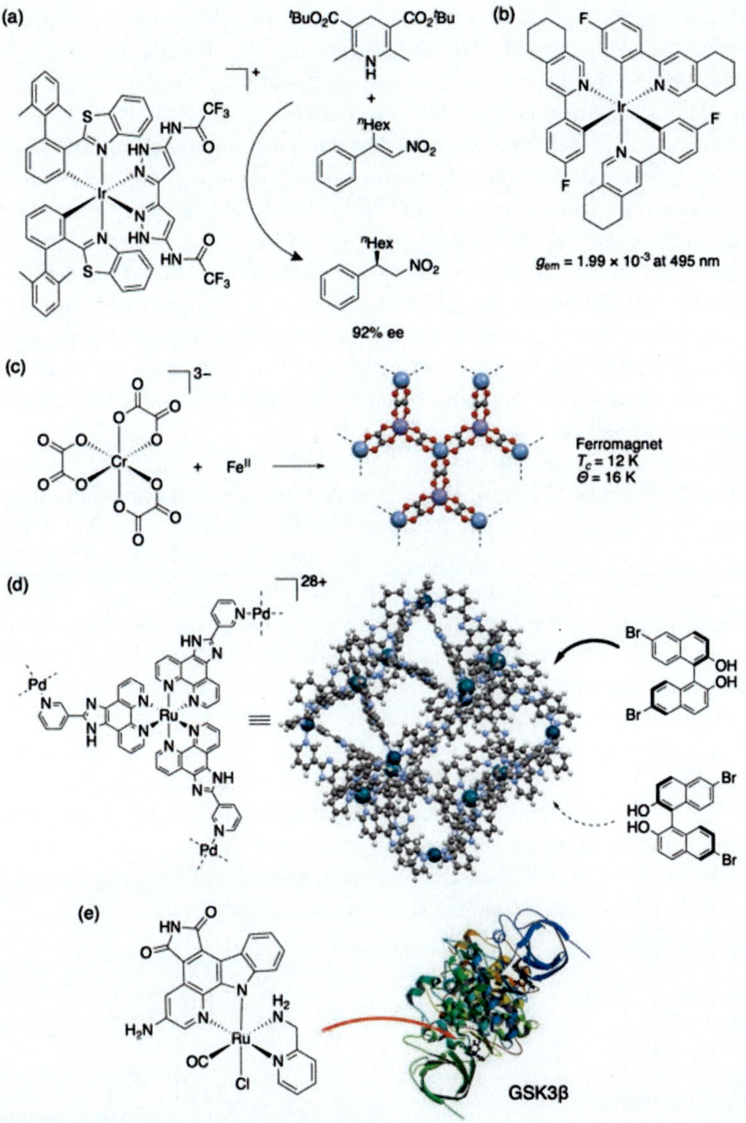

Fig. 3.5 Examples of applications of configurationally stable chiral metal centers: **a** enantioselective catalysis [36], **b** circularly polarized luminescence [37], **c** a chiral magnetic material [38], **d** enantioseparation [39], and **e** enzyme inhibition [40]

[33–35], tetrahedral ones have never been utilized for such a purpose to the best of my knowledge. This is partly because all four ligands strongly are coordinated to the metal atom in the previously reported configurationally stable tetrahedral metal centers. This situation precludes the direct access of a substrate molecule to the metal atom, which significantly diminishes the potential opportunities of catalysis. Therefore, a tetrahedral metal center which acts both as a configurationally stable chirality source and as a substrate activation site should be developed to promote application to enantioselective catalysis.

If these problems are amply resolved, configurationally stable tetrahedral chiral metal centers will be able to be widely used like octahedral and half-sandwich ones. Since tetrahedral metal centers have different steric properties, ligand exchange kinetics, and catalytic activities from octahedral and half-sandwich metal centers, such expansion of the scope would be significant and fruitful for the chemistry of chiral metal complexes.

Herein, I present a chiral-at-metal Zn^{II} complex with a configurationally stable tetrahedral Zn^{II} center (Fig. 3.6). This complex is highly configurationally stable, can be synthesized enantioselectively, and acts as an enantioselective catalyst. The design of this complex is based on an unsymmetrical tridentate ligand with a strong and non-planar coordination ability. This ligand rendered the tetrahedral Zn^{II} center highly stable against stereoinversion even at high temperatures, even though it is a mononuclear high-valent metal center. This ligand also provided one labile coordination site. Using this property, the enantiomeric forms of this complex were enantioselectively obtained by connecting a chiral ligand temporarily. Furthermore, this labile coordination site also worked as a substrate activation site in a catalytic enantioselective oxa-Diels–Alder reaction. These results reveal the previously underestimated potentials of tetrahedral metal centers, which can integrate chirality with metal-based functions without any other chiral components. Also, the increased availability of configurationally stable tetrahedral chiral metal centers will bring new opportunities to every field of chemistry which deals with chiral molecules.

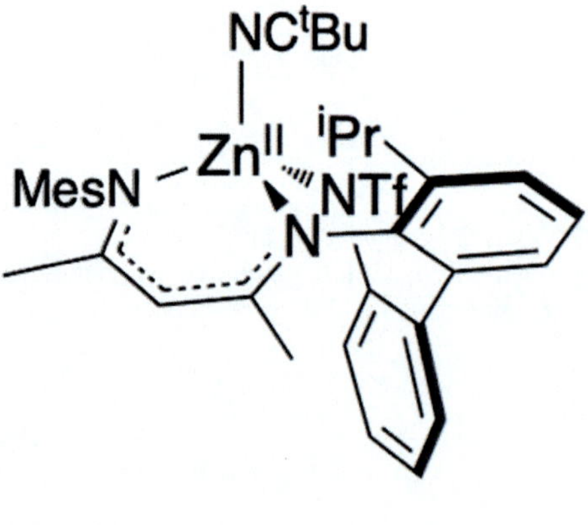

Fig. 3.6 Chiral-at-metal Zn^{II} complex with a configurationally stable tetrahedral Zn^{II} center developed in this study

The enantioselective synthesis of this complex was achieved by a kinetically controlled stepwise method. Since a pair of enantiomers have the same thermodynamic energy, kinetic control was necessary to gain enantioselectivity. In this work, enantioselective synthesis was achieved by three steps: racemic complexation of the tridentate ligand with metal, asymmetric induction by a chiral ligand, and replacement of the chiral ligand by an achiral ligand. The last step was conducted under kinetic control with support of the aforementioned property of the tridentate ligand, which kinetically stabilizes the absolute configuration of the metal center. Meanwhile, the second step was conducted under thermodynamic control by switching the stereoinversion kinetics at the metal center using the weakly acidic chiral ligand. This synthetic procedure would be useful for general chiral metal centers with not only tetrahedral geometry but also other geometry.

3.2 Molecular Design

To develop a useful configurationally stable chiral tetrahedral metal center, I set out from design of a new ligand. In contrast to the previous researches, I thought that it may be possible to stabilize the configuration of a mononuclear tetrahedral metal center in any d-electron state with a labile coordination site if an appropriate ligand was employed.

A tridentate ligand seemed suitable for this purpose because it can maximize its interactions with a metal atom while leaving one labile coordination site. To make a metal center chiral, the tridentate ligand should be unsymmetrical with three different coordination sites (Fig. 3.7a).

When such a ligand is used, there are two possible types of pathways of stereoinversion: dissociation and planarization. In the dissociation pathways, the tridentate ligand dissociates from the metal center partially or completely to lose the stereoinformation (Fig. 3.7b). In the planarization pathways, the tridentate ligand becomes planar without dissociation, during associative or dissociative ligand exchange of the

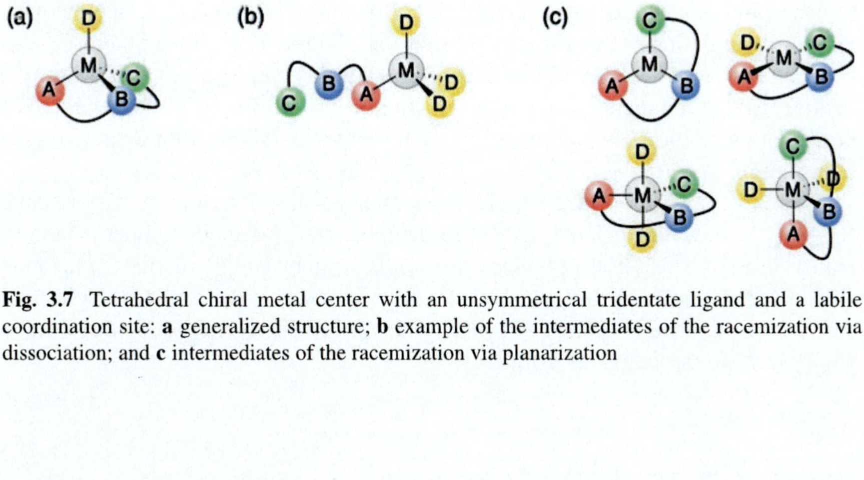

Fig. 3.7 Tetrahedral chiral metal center with an unsymmetrical tridentate ligand and a labile coordination site: **a** generalized structure; **b** example of the intermediates of the racemization via dissociation; and **c** intermediates of the racemization via planarization

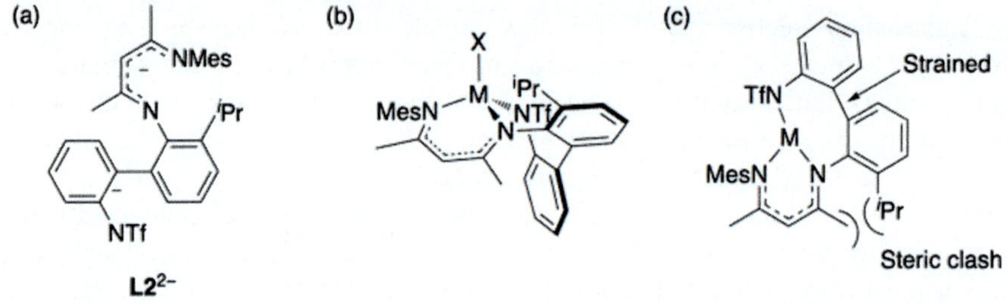

Fig. 3.8 Ligand **L2**, designed to stabilize the absolute configuration of a tetrahedral metal center with a labile coordination site: **a** structure in a dianionic state; **b** expected structure of its complex with metal M and labile ligand X; and **c** hypothetical planarized state of its complex showing a steric clash and strain

labile ligand, or by polytopal isomerization into square-planar geometry, and thereby the stereoinformation will be lost (Fig. 3.7c). This consideration suggests that the configuration of a tetrahedral metal center can be stabilized if both of these two kinds of pathways are effectively blocked.

With this view, I designed tridentate ligand **L2** (Fig. 3.8a, b). Firstly, the donor atoms of this ligand are negatively charged, which will be electrostatically attracted to a metal cation in nonpolar media. Secondly, the donor atoms are amido nitrogens, which have high Lewis basicity. Thirdly, the β-diketiminate moiety offers a very rigid chelation, while its pendant arm also forms a chelate ring. These three features offer strong coordination of this ligand to a metal atom and therefore may rule out the dissociation pathways. Meanwhile, if this ligand is planarized while coordinating, the isopropyl group will cause a steric clash with the closer methyl group, and also the biaryl moiety will suffer from a great strain (Fig. 3.8c) [55]. Thus, this ligand precludes the planarization pathways as well. In this way, this ligand is expected to effectively prevent rapid racemization. It should be noted that this ligand is flexible and thus achiral in a free state. The geometry of this ligand is suited for a tetrahedral metal center, while the steric shielding prohibits coordination of an additional ligand.

In the initial stage of this research, a similar ligand with an *N*,*O*-coordinating β-ketoiminate moiety instead of the β-diketiminate moiety and a *p*-toluenesulfonamidate moiety instead of the triflamidate moiety had been employed. However, preliminary experiments showed that the configurational stability of its Zn^{II} complex was low. Thereafter, β-diketiminate was employed to afford stronger coordination to the metal center, and triflamidate was employed to further enhance coordination of the β-diketiminate moiety by reducing the electron density on the metal atom.

The mesityl group of **L2** was selected after an initial screening with phenyl, 2,6-xylyl, 2,6-dichlorophenyl, 2,4,6-tribromophenyl, 2,6-dimethoxyphenyl, and 9-anthryl groups. This series of ligands was synthesized by simply modifying the final step of the synthesis. Among them, **L2** with the mesityl group showed the highest selectivity in the enantioselective synthesis, probably due to the steric effects and a strong OH–π interaction (*vide infra*).

3.3 Strategy for Enantioselective Synthesis

I planned enantioselective construction of this tetrahedral metal center using a chiral auxiliary. Since a pair of enantiomers always have the same thermodynamic energy, enantioselectivity can be realized only under kinetic control, and a stepwise method is the easiest way to achieve a kinetically controlled synthesis. Previously, enantioselective construction of configurationally stable octahedral metal centers has been conducted by stepwise methods using chiral ligands as chiral auxiliaries [56]. Inspired by these reports, I envisioned the use of a chiral ligand as a chiral auxiliary which can be coordinated to the labile coordination site of the metal–**L2** complex. In this case, the enantioselective construction can be conducted in three steps (Fig. 3.9): (1) Complexation of ligand **L2** and metal species is conducted to generate a racemic mixture of complexes; (2) asymmetric induction at the metal center is conducted by addition of a chiral ligand, which biases diastereomers in equilibrium; (3) the chiral ligand is replaced by an achiral ligand without changing the configuration of the metal center.

There are two possible difficulties in this method. Firstly, the last step needs to be kinetically controlled to finish ligand exchange before stereoinversion of the metal center happens. This is possible if the absolute configuration of the metal center was amply stabilized by ligand **L2** by the aforementioned properties. Secondly, in turn, the second step needs to be thermodynamically controlled to give a high selectivity in the absolute configuration of the metal center. If the metal center with **L2** is highly configurationally stable, this is difficult to achieve. As mentioned later, I employed a weakly acidic chiral ligand to solve this problem by temporarily weakening the coordination of **L2** to the metal atom.

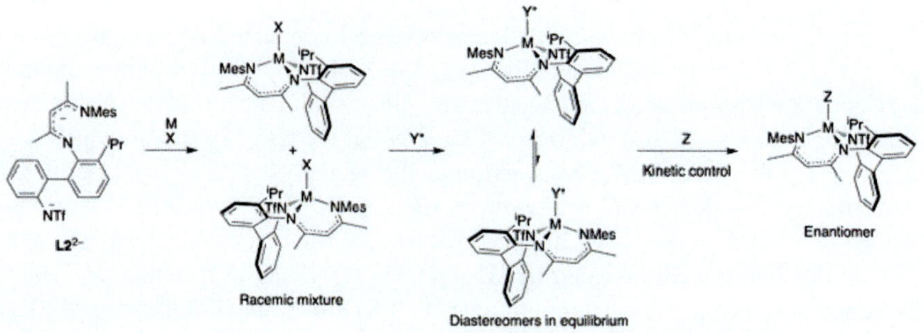

Fig. 3.9 Reaction scheme planned for enantioselective construction of a tetrahedral metal center with **L2** using a chiral ligand as a chiral auxiliary. M, metal species; X, an achiral ligand; Y*, a chiral ligand; and Z, an achiral ligand

3.4 Step 1: Racemic Complexation

Ligand H_2**L2** was synthesized in 5 steps from commercially available starting materials. I chose Zn^{II} as a metal center because the net charge of its complex will be neutral with **L2**$^{2-}$ and a neutral monodentate ligand, which may guarantee high solubility in nonpolar solvents as well as the lability of the monodentate ligand. A tetrahedral Zn^{II} center is also interesting in relation to biological systems because there are many metalloenzymes which contain a tetrahedral Zn^{II} center [52], and because some of them even work as enantioselective catalysts due to the chirality of the protein backbone [53].

For the first step of enantioselective synthesis of a Zn^{II} complex, I conducted racemic complexation. Diethylzinc was employed as a metal source which can deprotonate H_2**L2** upon complexation (Fig. 3.10). The ^{1}H and ^{19}F NMR spectra showed clean formation of one species in 98% yield as estimated from ^{1}H NMR spectroscopy using an internal standard. All the ^{1}H signals were assigned by ^{1}H–^{1}H COSY and NOESY techniques, and it was confirmed that ligand H_2**L2** was doubly deprotonated. Single-crystal XRD analysis showed the solid-state structure of this product to be the homochiral dimeric complex (*R**,*R**)-[Zn_2**L2**$_2$] where the sulfonyl oxygen of another {Zn**L2**} unit acted as a labile monodentate ligand. The pattern of NOE signals in the ^{1}H–^{1}H NOESY spectrum accorded with this crystal structure, indicating that this structure is mostly maintained in solution state as well. This structure confirmed the expected tetrahedral geometry of the Zn^{II} center.

3.5 Step 2: Asymmetric Induction with a Chiral Ligand

For the second step of enantioselective synthesis, I conducted asymmetric induction at the Zn^{II} center using a chiral ligand. (*S*)-α,α-Diphenyl-2-pyrrolidinemethanol ((*S*)-dpp) was chosen as a chiral ligand. This ligand can be coordinated to Zn^{II} via the nitrogen atom and also exhibits weak acidity at the hydroxy group, which was necessary to facilitate the stereoinversion of the {Zn**L2**} center as discussed later. Accordingly, the aforementioned complex (*R**,*R**)-[Zn_2**L2**$_2$] was treated with (*S*)-dpp (Fig. 3.11). ^{1}H and ^{19}F NMR spectra showed formation of one major product. All the ^{1}H signals were assigned by ^{1}H–^{1}H COSY and NOESY techniques, and it was confirmed that (*S*)-dpp was coordinated to the {Zn**L2**} unit. Single-crystal XRD analysis showed that this product was the diastereomer (*S*)-[Zn**L2**((*S*)-dpp)]. The pattern of NOE signals in the ^{1}H–^{1}H NOESY spectrum accorded with this crystal structure, indicating that the crystal structure is mostly maintained in solution state as well. This structure confirmed successful asymmetric induction.

The crystal structure showed three non-covalent interactions: an NH–O hydrogen bond (2.870 Å) between the (*S*)-dpp and triflyl moieties, an NH–O hydrogen bond (2.719 Å) within the (*S*)-dpp moiety, an OH–π interaction (3.180 Å) between the (*S*)-dpp and mesityl moieties (Fig. 3.11b). These non-covalent interactions and the

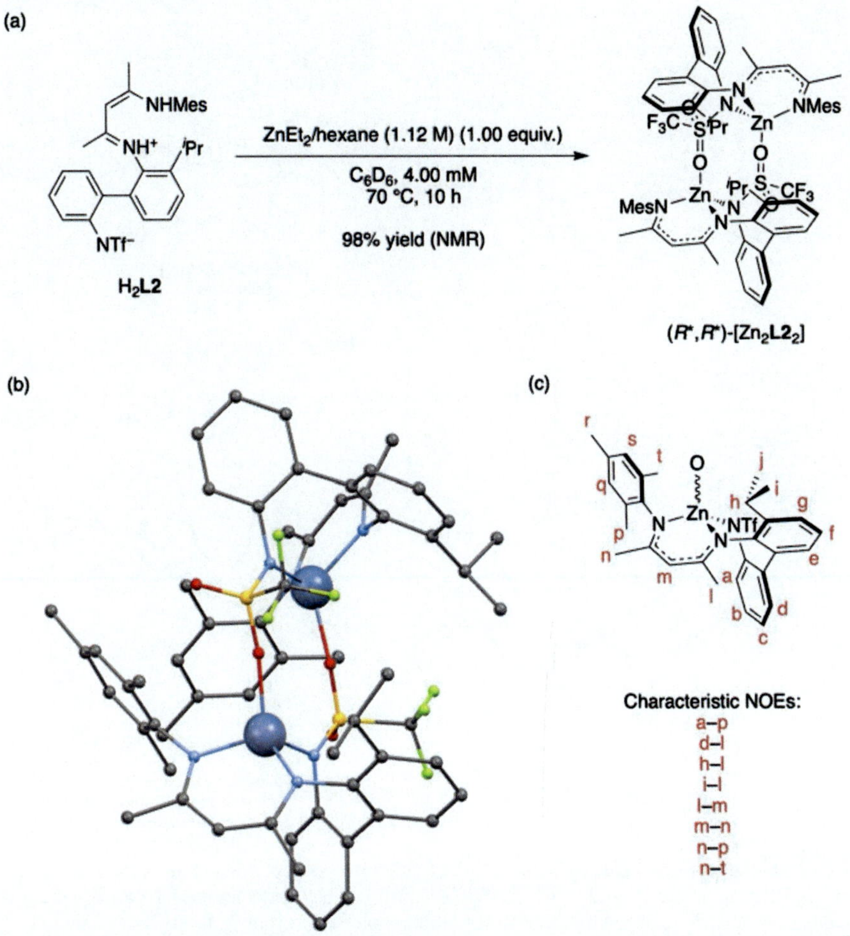

Fig. 3.10 Synthesis of racemic complex (R*,R*)-[Zn_2**L2**$_2$]. **a** Reaction scheme. **b** Single-crystal XRD structure in a ball-and-stick model. The hydrogen atoms are omitted for clarity. Color code: Zn, blue-gray; C, gray; N, blue; O, red; F, yellow-green; and S, yellow. **c** Labeling of the hydrogen atoms and list of the characteristic NOEs

steric repulsion between the (S)-dpp and **L2** backbones would be the cause of the biased diastereomer equilibrium.

To investigate the process of the asymmetric induction, I conducted a time-course NMR analysis. The NMR spectra right after addition of (S)-dpp showed formation of (S)-[Zn**L2**((S)-dpp)] as a major species, a second major species, and one minor species (Fig. 3.12). The second major species can be assigned to the other diastereomer (R)-[Zn**L2**((S)-dpp)], judging from the NMR pattern similar to that of (S)-[Zn**L2**((S)-dpp)]. The signals of the minor species could not be fully assigned, but the presence of a doublet signal with a large coupling constant in the ^{1}H NMR spectrum suggested protonation at the β-diketiminate moiety of **L2**. The source of

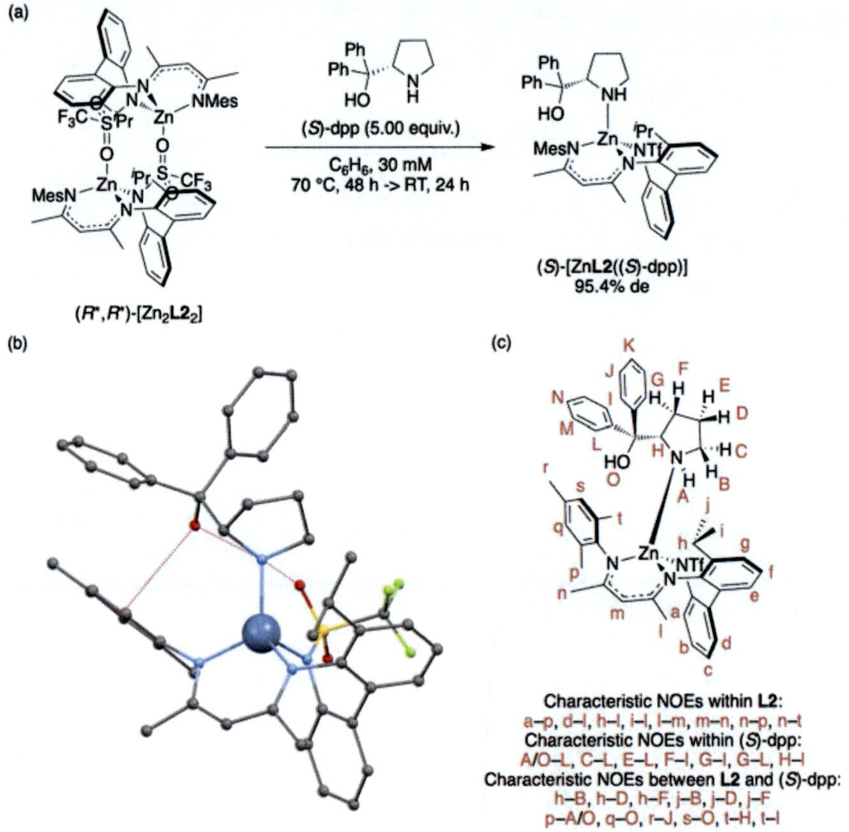

Fig. 3.11 Asymmetric induction at the {Zn**L2**} unit with (*S*)-dpp. **a** Reaction scheme. **b** Single-crystal XRD structure of product (*S*)-[Zn**L2**((*S*)-dpp)] in a ball-and-stick model. One of the crystallographically independent but structurally similar molecules is shown. The hydrogen atoms and solvent molecules are omitted for clarity. The hydrogen bonds and an OH–π interaction are drawn in dashed magenta lines. Color code: Zn, blue-gray; C, gray; N, blue; O, red; F, yellow-green; and S, yellow. **c** Labeling of the hydrogen atoms and list of the characteristic NOEs

proton would be the hydroxy group of coordinated (*S*)-dpp, which means this minor species may be [Zn(H**L2**)(H_{-1}dpp)]. The initial ratio of these three species was 54:42:4, which can be assigned to (*S*)-[Zn**L2**((*S*)-dpp)], (*R*)-[Zn**L2**((*S*)-dpp)], and (*R*)-[Zn(H**L2**)((*S*)-H_{-1}dpp)], respectively, generated from a 1:1 mixture of (*R*,*R*)-[Zn_2**L2**$_2$] and (*S*,*S*)-[Zn_2**L2**$_2$] with a minute change in the absolute configuration at the Zn^{II} centers. On the other hand, when the mixture was heated at 70 °C for 48 h, the ratio dramatically changed into 96:4:1, showing that the asymmetric induction proceeded via stereoinversion at the {Zn**L2**} center.

While the stereoinversion was beneficial for the asymmetric induction, it should be avoided after the removal of (*S*)-dpp to afford configurational stability in the final product. I expected that the stereoinversion of the {Zn**L2**} unit can be accelerated

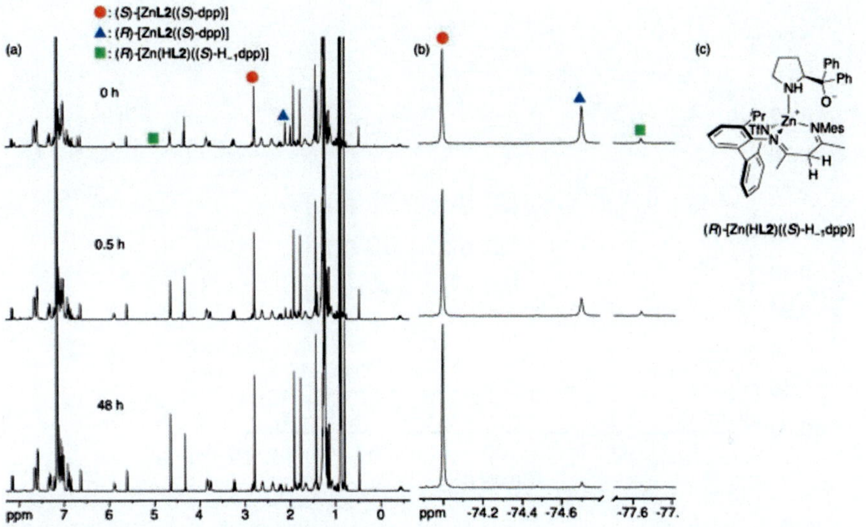

Fig. 3.12 Time-course NMR analysis for the asymmetric induction at the {Zn**L2**} unit with (*S*)-dpp. Reaction conditions: (*R**,*R**)-[Zn_2**L2**$_2$] + (*S*)-dpp (2.50 equiv.), MS 4A, C_6D_6, 4.00 mM, 70 °C. **a** ^{1}H NMR spectra (C_6D_6, 300 K, 500 MHz). **b** ^{19}F NMR spectra (C_6D_6, 300 K, 471 MHz). **c** Plausible structure of (*R*)-[Zn(H**L2**)((*S*)-H_{-1}dpp)]

only in the presence of (*S*)-dpp, possibly by proton transfer via (*R*)-[Zn(H**L2**)((*S*)-H_{-1}dpp)]. To confirm this, I examined asymmetric induction with a similar chiral ligand, (*S*)-2-(methoxydiphenylmethyl)pyrrolidine ((*S*)-mdp), which has a methoxy group instead of the hydroxy group in (*S*)-dpp (Fig. 3.13). In this case, the diastereomer ratio during heating stayed approximately at 1:1, even though there should be some energy difference between diastereomers. This result suggests that the presence of a hydroxy group in (*S*)-dpp accelerated stereoinversion of the Zn^{II} center exactly as expected. As another experiment, I also examined the effects of the amount of (*S*)-dpp on the kinetics of the asymmetric induction. When only 1 equiv. of (*S*)-dpp was used instead of 2.5 equiv., the asymmetric induction was significantly decelerated. These results indicate that the observed stereoinversion kinetics during the asymmetric induction with (*S*)-dpp was accelerated by the presence of 2.5 equiv. of (*S*)-dpp, and the configurational stability may become sufficient once (*S*)-dpp is removed.

3.6 Step 3: Replacement of the Chiral Ligand

For the third step of enantioselective synthesis, I conducted replacement of the chiral ligand (*S*)-dpp with an achiral ligand. I employed ^{t}BuCN as an achiral ligand, expecting that its moderate coordination affinity to Zn^{II} leads to smooth replacement

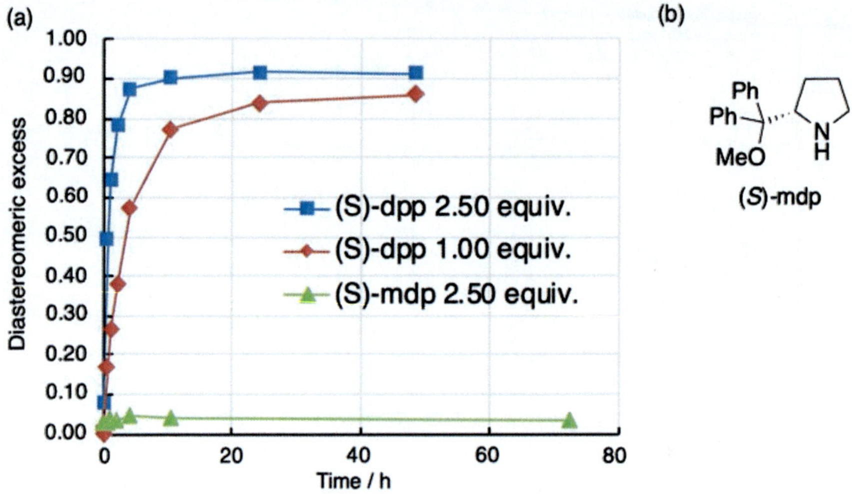

Fig. 3.13 **a** Kinetics of the asymmetric induction at the {Zn**L2**} unit under various conditions. Reaction conditions: (R^*,R^*)-[Zn_2**L2**$_2$] + (S)-dpp or (S)-mdp (2.50 or 1.00 equiv.), dry MS 4A, C_6D_6, 4.00 mM, 70 °C. The diastereomeric excess was calculated from the integral ratios of diastereomers including (R)-[Zn(H**L2**)((S)-H_{-1}dpp)] in the ^{19}F NMR spectra. **b** Structure of (S)-mdp

of (S)-dpp, yet afterward it can act as a labile ligand. Accordingly, the aforementioned diastereoenriched Zn**L2**–(S)-dpp complex was treated with a large excess amount of ^{t}BuCN, and the product was directly crystallized by addition of a poor solvent (Fig. 3.14). ^{1}H and ^{19}F NMR spectra suggested the isolation of a new product. Assignment of proton signals by ^{1}H–^{1}H COSY and NOESY techniques and elemental analysis confirmed that (S)-dpp coordinated to the {Zn**L2**} unit was replaced by ^{t}BuCN. Single-crystal XRD analysis showed the molecular structure of [Zn**L2**(^{t}BuCN)]. The pattern of NOE signals in the ^{1}H–^{1}H NOESY spectrum accorded with this crystal structure, indicating that the crystal structure is maintained in solution state as well. In this way, the chiral ligand was successfully replaced to furnish a chiral-at-metal complex. The total yield from H_2**L2** was 72%.

The enantiopurity of the product was determined by ^{19}F NMR spectroscopy with a chiral shift reagent, (R)-methyl p-tolyl sulfoxide ((R)-mts) (Fig. 3.15). First, a mixture of racemic (R^*,R^*)-[Zn_2**L2**$_2$] and (R)-mts was examined by ^{19}F NMR spectroscopy for comparison. The spectrum showed two separated signals in a 50:50 integral ratio, which would correspond to (R)-[Zn**L2**((R)-mts)] and (S)-[Zn**L2**((R)-mts)] formed by coordination of the oxygen atom of (R)-mts to the Zn^{II} center. This result verified the accuracy of this method for enantiopurity determination. In contrast, the ^{19}F NMR spectrum of a mixture of the aforementioned sample of [Zn**L2**(NC^{t}Bu)] and (R)-mts showed only one signals with a signal-to-noise ratio of 1282, corresponding to > 99.5% ee assuming the detection limit in the signal-to-noise ratio as 3. The major absolute configuration was determined to be S from the single-crystal XRD data with a Flack parameter of −0.002(5). These results indicate that the chiral

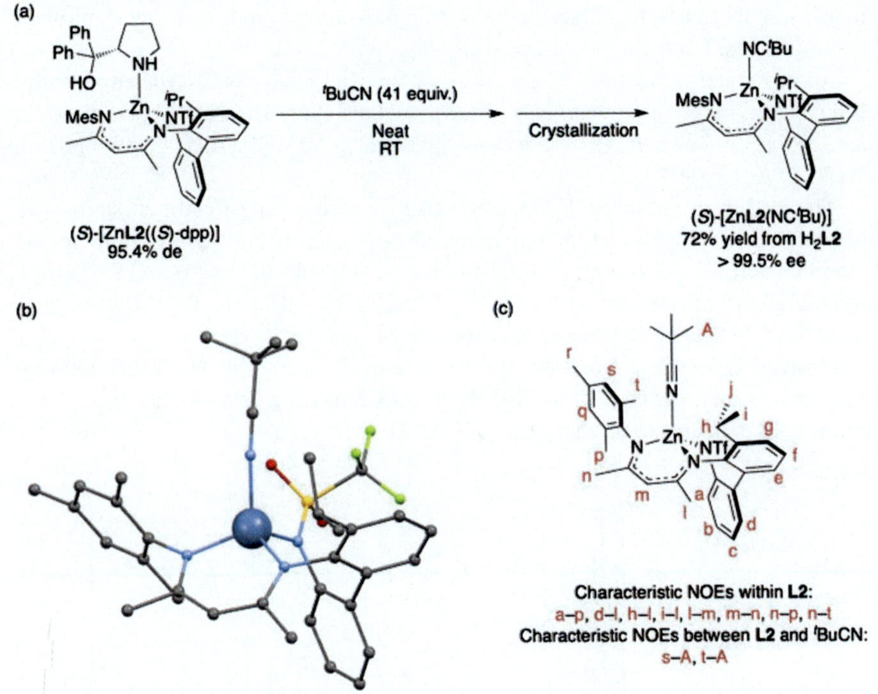

Fig. 3.14 Removal of chiral auxiliary from the Zn**L2**–(*S*)-dpp complexes. **a** Reaction scheme. **b** Single-crystal XRD structure of product (*S*)-[Zn**L2**(NC^{t}Bu)] in a ball-and-stick model. The hydrogen atoms are omitted for clarity. Color code: Zn, blue-gray; C, gray; N, blue; O, red; F, yellow-green; and S, yellow. **c** Labeling of the hydrogen atoms and list of the characteristic NOEs

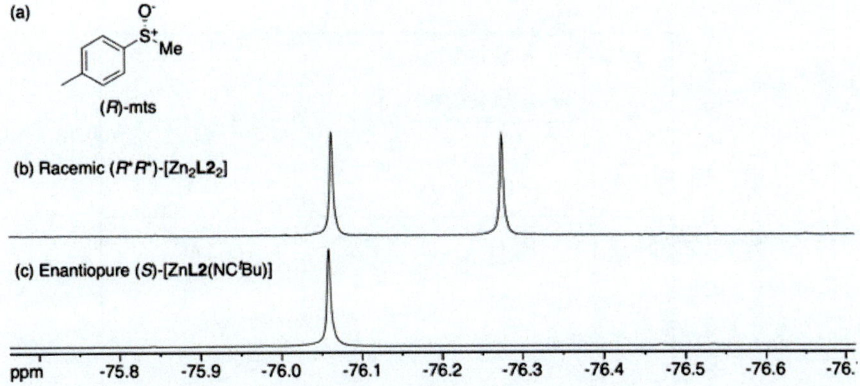

Fig. 3.15 Enantiopurity determination of [Zn**L2**(NC^{t}Bu)]. **a** Structure of chiral shift reagent (*R*)-mts. **b**, **c** ^{19}F NMR spectra (C_6D_6, 300 K, 471 MHz): **a** racemic (R^*R^*)-[Zn_2**L2**$_2$] + (*R*)-mts (1.10 equiv.); **b** enantioselectively synthesized [Zn**L2**(NC^{t}Bu)] + (*R*)-mts (6 equiv.)

ligand was successfully replaced with retention of chirality to furnish an enantiopure chiral-at-metal complex.

In the crystal structure, the geometry of the Zn^{II} center was tetrahedral with a little distortion, giving a geometry index $\tau_4 = 0.85$ [56]. The Zn–N bond distances are 1.961(2) Å (mesityl N), 1.955(2) Å (middle N in **L2**), 2.017(2) Å (triflyl N), and 2.029(2) Å (nitrile N).

The used chiral auxiliary (*S*)-dpp was able to be recovered. After the crystallization of (*S*)-[Zn**L2**(NC^{t}Bu)], the (*S*)-dpp in the supernatant was purified by aminopropyl-modified silica gel column chromatography. As a result, 95% of pure (*S*)-dpp was recovered as confirmed by ^{1}H NMR (Fig. 3.16). This result shows the chiral auxiliary in this enantioselective synthesis method can be recycled.

(*R*)-[Zn**L2**(NC^{t}Bu)] was also synthesized in the same manner using (*R*)-dpp. The circular dichroism (CD) spectra of the [Zn**L2**(NC^{t}Bu)] enantiomers showed an expected mirror-image relationship (Fig. 3.17).

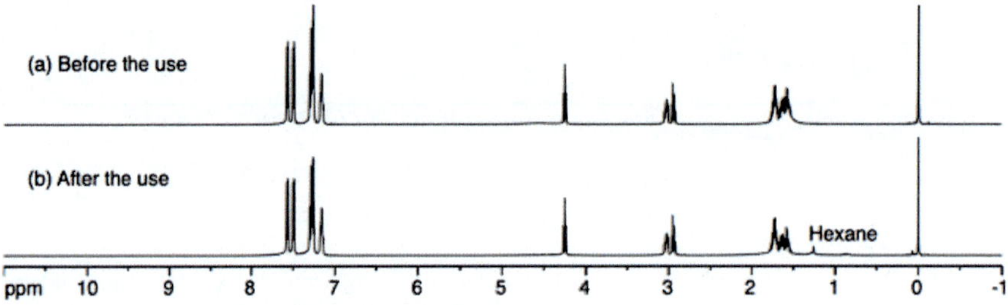

Fig. 3.16 ^{1}H NMR spectra ($CDCl_3$, 300 K, 500 MHz) of (*S*)-dpp **a** before the use and **b** after the use

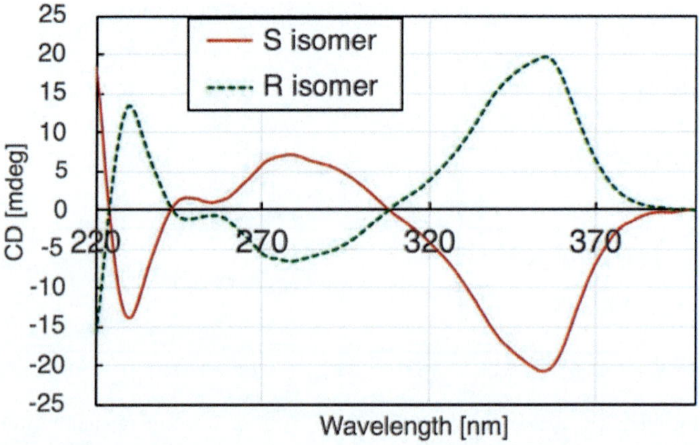

Fig. 3.17 CD spectra of the enantiomers of [Zn**L2**(NC^{t}Bu)] (1,2-dichloroethane, 0.19 mM, l = 0.10 cm)

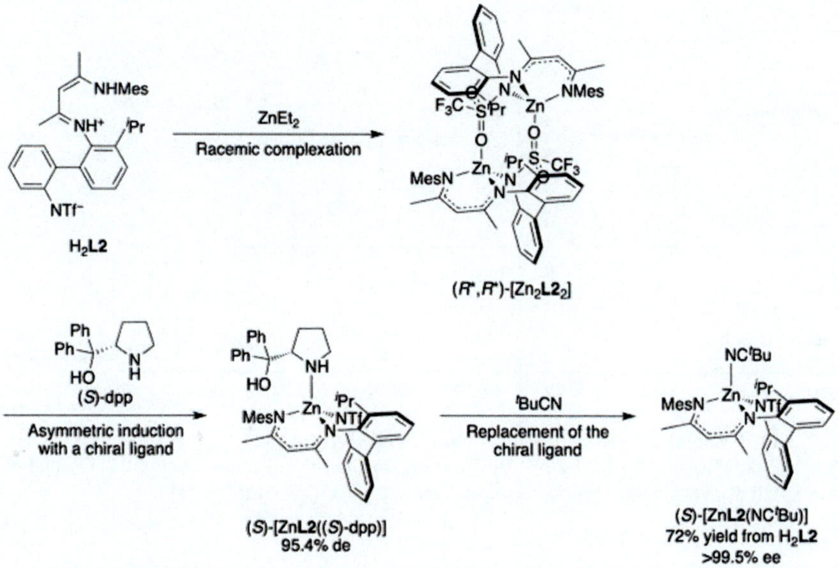

Fig. 3.18 Overall reaction scheme of the enantioselective synthesis of (*S*)-[Zn**L2**(NC^{t}Bu)]

In this way, enantioselective synthesis of (*S*)-[Zn**L2**(NC^{t}Bu)] was accomplished in three steps (Fig. 3.18). To the best of my knowledge, such multistep enantioselective construction of a metal center using a chiral ligand as a chiral auxiliary was previously known only in the case of octahedral metal centers without labile coordination sites [56]. The present work is the first example both for construction of tetrahedral metal centers and for construction of metal centers with a labile coordination site, which has high potential as a substrate activation site.

3.7 Configurational Stability

With a pure enantiomer in hand, I investigated the configurational stability of complex [Zn**L2**(NC^{t}Bu)]. (*S*)-[Zn**L2**(NC^{t}Bu)] was dissolved in C_6D_6 and left at 24 °C or 70 °C for 24 h, and then the enantiopurity was determined using (*R*)-mts (Fig. 3.19). At 24 °C, the decrease in enantiomeric excess was observed in 0.02(4)%. Even at 70 °C, the decrease was only 0.35(4)%. These results established the high configurational stability of this complex, even at a high temperature.

This configurational stability is remarkable as a mononuclear tetrahedral Zn^{II} complex. For comparison, typical mononuclear tetrahedral neutral Zn^{II} complexes are reported to show stereoinversion within the timescale of ^{1}H NMR measurement at room temperature in $CDCl_3$ [57, 58]. In the present system, ligand **L2** and the

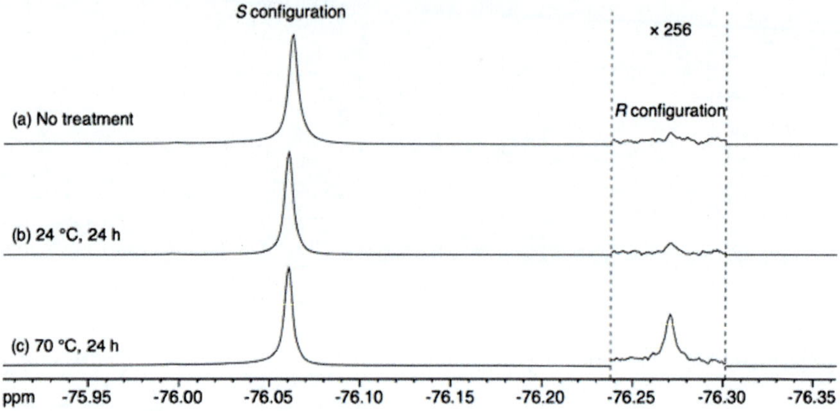

Fig. 3.19 Configurational stability test of [Zn**L2**(NC^{t}Bu)] in C_6D_6. ^{19}F NMR spectra (C_6D_6, 300 K, 500 MHz) of (*S*)-[Zn**L2**(NC^{t}Bu)] treated with (*R*)-mts **a** immediately after dissolution; **b** after left at 24 °C, 4.0 mM for 24 h; and **c** after left at 70 °C, 4.0 mM for 24 h

solvent condition (dehydrated C_6D_6) effectively stabilized the configuration of the Zn^{II} center.

The coordination of ^{t}BuCN to this complex may affect the configurational stability of the {Zn**L2**} center. As a preliminary experiment, the effects of coordination of ^{i}PrOH were examined. When 2.5 equiv. of ^{i}PrOH was added to (*S*)-[Zn**L2**(NC^{t}Bu)] in C_6D_6, the ^{19}F NMR spectrum showed two signals with comparative integral values, indicating partial formation of (*S*)-[Zn**L2**(HO^{i}Pr)]. Heating this solution at 70 °C for 24 h resulted in the decrease in the enantiomeric excess by 0.09(7)%, as confirmed by addition of (*R*)-mts. This decrease is smaller than that of (*S*)-[Zn**L2**(NC^{t}Bu)] alone, which indicates the coordination of ^{i}PrOH and/or the presence of excess ^{i}PrOH further stabilizes the absolute configuration of Zn^{II} center.

3.8 Enantioselective Catalysis

It was expected that the ^{t}BuCN site of the complex [Zn**L2**(NC^{t}Bu)] can work as a labile coordination site. To show the potential utility of this labile coordination site, enantioselective catalysis was conducted using this complex as a chiral Lewis acid. As a model reaction, a typical oxa-Diels–Alder reaction was chosen [59].

Reaction of Danishefsky's diene **8** and aldehyde **9** were conducted with a catalytic amount of (*S*)-[Zn**L2**(NC^{t}Bu)] (Fig. 3.20). After acidic work-up, the product was analyzed by ^{1}H NMR spectroscopy and chiral HPLC, which indicated that (*R*)-**10** was obtained in 78% yield and 88% ee. In the absence of the catalyst, no reactions were confirmed. These results clearly show that the complex (*S*)-[Zn**L2**(NC^{t}Bu)] acted as a chiral Lewis acid.

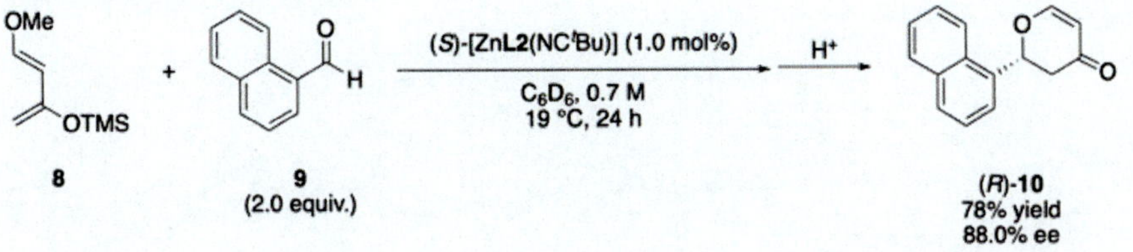

Fig. 3.20 Reaction scheme of the enantioselective oxa-Diels–Alder reaction catalyzed by (*S*)-[Zn**L2**(NC^{t}Bu)]

To investigate the mechanism, I prepared the substrate complex [Zn**L2**(**9**)] from (*R**,*R**)-[Zn_2**L2**$_2$] and aldehyde **9** (Fig. 3.21). The ^{1}H and ^{19}F NMR spectra showed

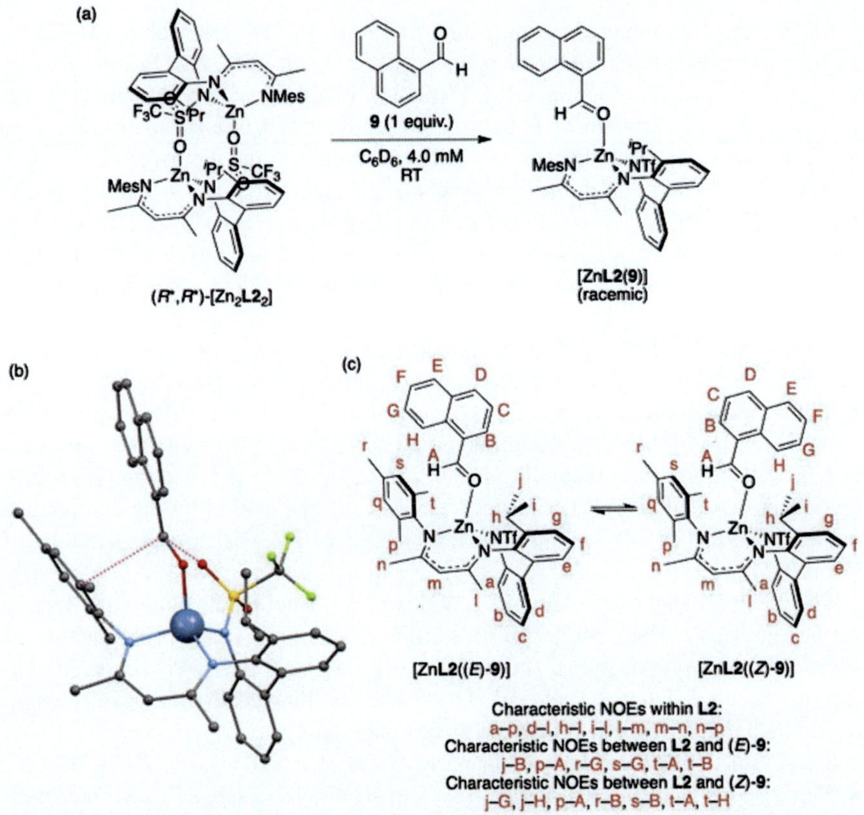

Fig. 3.21 Synthesis of [Zn**L2**(**9**)]. **a** Reaction scheme. **b** Single-crystal XRD structure in a ball-and-stick model. The *R* enantiomer is shown. The hydrogen atoms are omitted for clarity. The hydrogen bond and a π–π interaction are drawn in dashed magenta lines. Color code: Zn, blue-gray; C, gray; N, blue; O, red; F, yellow-green; and S, yellow. **c** Labeling of the hydrogen atoms and list of the characteristic NOEs for the diastereomers regarding the flipping of naphthyl group in equilibrium

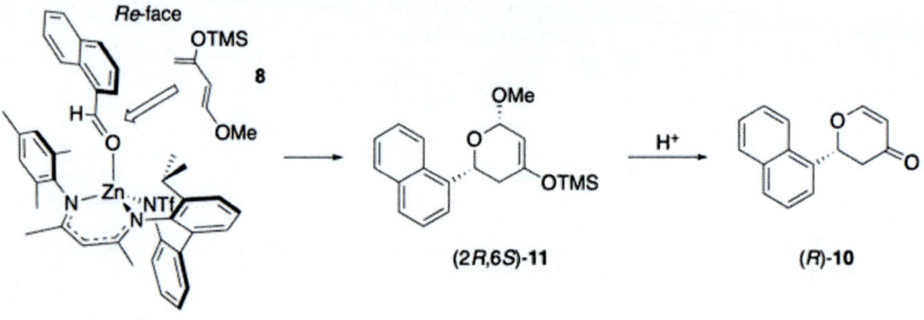

Fig. 3.22 Plausible mechanism of the enantioselective formation of (*R*)-**10**. The preferred attacking side of (*R*)-[Zn**L2**(**9**)]

formation of one product. Assignment of ^{1}H signals by ^{1}H–^{1}H COSY and NOESY techniques revealed that aldehyde **9** was coordinated to the {Zn**L2**} unit. Single-crystal XRD analysis determined the structure of this complex. Aldehyde **9** was coordinated to the Zn^{II} atom via the oxygen atom, and its orientation was fixed by a CH–O hydrogen bond between the aldehyde hydrogen and the triflamide oxygen of **L2** (2.915 Å) and a π–π interaction between the aldehyde carbon and the mesityl group of **L2** (3.097 Å). The pattern of NOE signals in the ^{1}H–^{1}H NOESY spectrum accorded with this crystal structure when flipping of the naphthyl group was taken into account. This indicates that the crystal structure was maintained in solution state as well.

The formation of (*R*)-[Zn**L2**(**9**)] (the change in the stereochemical descriptor is due to the high priority of **9** in the Cahn–Ingold–Prelog rule) under the catalytic condition was confirmed from ^{19}F NMR measurements. At the initial stage, the ^{19}F NMR spectrum of the reaction mixture showed a signal at the same chemical shift as [Zn**L2**(**9**)], indicating that ^{t}BuCN was replaced by **9**. As the reaction proceeded, the signal was shifted and broadened, which would be caused by the competitive coordination of (2*R*,6*S*)-**11** in the fast exchange (see Fig. 3.22 for the structure of (2*R*,6*S*)-**11**).

The Lewis acidity of the {Zn**L2**} center was estimated by the Gutmann–Beckett method [60, 61]. The complex [Zn**L2**(OPEt$_3$)] was prepared by the reaction of (*R**,*R**)-[Zn$_2$**L2**$_2$] and Et$_3$PO, and its ^{31}P NMR chemical shift was compared with free Et$_3$PO. In C$_6$D$_6$, the chemical shift of the coordinated Et$_3$PO was 21.1 ppm higher than that of free Et$_3$PO. Given the acceptor number (*AN*) of solvent benzene is 8.2, *AN* of {Zn**L2**} can be calculated by $AN = 21.1 * 100/(86.14 - 41.0) + 8.2 = 55$ [62]. This value is comparable to SnCl$_4$ ($AN = 59$), which is a Lewis acid with moderate strength [63].

In the complex (*R*)-[Zn**L2**(**9**)], the Lewis acidity of {Zn**L2**} would activate the coordinated **9** for the nucleophilic attack by **8**. In the structure of (*R*)-[Zn**L2**(**9**)], the *Si*-face of **9** is effectively shielded by the mesityl group of **L2**, so the attack of **8** would preferentially occur from the *Re*-face (Fig. 3.22). This attack generated

(2*R*,6*S*)-**11**, which was supported by ^{1}H NMR monitoring. (2*R*,6*S*)-**11** was converted into (*R*)-**10** after the acidic work-up.

The configurational stability of the {Zn**L2**} center during the catalysis was confirmed by the ^{19}F NMR measurement. When excess (*R*)-mts was added before the acidic work-up, the ^{19}F NMR spectrum showed only a signal of (*R*)-[Zn**L2**((*R*)-mts)] in a signal-to-noise ratio of 1282, which corresponds to > 99.5% ee. This result shows the {Zn**L2**} center is configurationally stable during the catalysis even though **9**, (2*R*,6*S*)-**11**, and ^{t}BuCN were dynamically exchanged as labile ligands.

In this way, the labile coordination site of the complex (*S*)-[Zn**L2**(NC^{t}Bu)] was shown to be useful in enantioselective catalysis. This result demonstrates the utility of the design of ligand **L2** which gives both configurational stability and the labile coordination site.

3.9 Conclusions

In this chapter, I presented a tetrahedral metal center which can be enantioselectively constructed, is configurationally stable, and provides labile coordination site available for enantioselective catalysis. Previously, chemistry of chiral molecules was mainly based on nonmetal centers or octahedral metal centers as configurationally stable chirality sources. The use of tetrahedral metal centers as configurationally stable chirality sources has been underdeveloped. In this work, I designed the unsymmetrical tridentate ligand **L2** with strong and non-planar coordination ability to stabilize the absolute configuration of a metal center, leaving one labile coordination site. The enantiopure complex [Zn**L2**(NC^{t}Bu)] was enantioselectively synthesized by temporarily attaching a chiral ligand to the labile site as a chiral auxiliary. The complex [Zn**L2**(NC^{t}Bu)] was shown to be remarkably stable against racemization. It was also demonstrated that this complex works as an enantioselective catalyst in oxa-Diels–Alder reaction by binding and activating a substrate at the labile coordination site. These results revealed the overlooked utility of tetrahedral chiral metal centers.

Enantioselective synthesis used here is an example of kinetically controlled stepwise synthesis. When the chiral ligand (*S*)-dpp was coordinated to the Zn^{II} center, the stereoinversion of the Zn^{II} center was accelerated and the system reached thermodynamic equilibrium. When ^{t}BuCN was added, fast ligand exchange occurred first, and then the stereoinversion of the Zn^{II} center became kinetically hindered. The design of **L2** which kinetically hinders stereoinversion in normal state but allows stereoinversion in the presence of a weakly acidic ligand played an important role in this multistep synthesis.

The product complex [Zn**L2**(NC^{t}Bu)] would be a good scaffold for preparation of complexes with attractive chiroptical properties such as circularly polarized luminescence. Since ^{t}BuCN in this complex is labile, various ligands can be easily introduced to this chiral Zn^{II} center. Introduction of fluorophore can lead to luminescence with a tunable color, while the {Zn**L2**} framework itself is luminescent in a blue color. In combination with chirality, circularly polarized luminescence may be possible.

3.10 Experimental Section

3.10.1 Materials and Methods

Unless otherwise noted, solvents and reagents were purchased from TCI Co., Ltd., FUJIFILM Wako Pure Chemical Corporation Ltd., Kanto Chemical Co., and Sigma-Aldrich Co., and used without further purification.

All the manipulations regarding the Zn^{II} complexes were conducted under dry N_2 atmospheres using a UNICO UN-650F glove box connected with a Glovebox Japan GBJPWS3 gas purifier, a UNICO UL-1000A glove box connected with a UNICO MT-1000X gas purifier, and gastight equipment, except for single-crystal XRD and work-up of catalysis. C_6H_6, C_6D_6, CD_2Cl_2, ^{t}BuCN, HMDSO, and 1,2-dichloroethane were dehydrated over MS 4A.

^{1}H, ^{13}C, ^{19}F, ^{11}B, and 2D NMR spectra were recorded on a Bruker AVANCE III-500 (500 MHz) spectrometer. Tetramethylsilane (TMS) was used as an internal standard (δ 0 ppm) for the ^{1}H and ^{13}C NMR measurements when $CDCl_3$ was used as a solvent. A residual solvent signal was used for calibration of the ^{1}H NMR measurements when C_6D_6 (δ 7.16 ppm), CD_3CN (δ 1.94 ppm), or CD_2Cl_2 (δ 5.32 ppm) was used as a solvent. No corrections were conducted for ^{19}F and ^{11}B NMR measurements. Abbreviations are as follows: s, singlet; d, doublet; t: triplet; br, broad; and m, multiplet.

ESI-MS data were recorded on a Micromass LCT Premier XE ESI-TOF mass spectrometer. The experimental conditions were as follows: desolvation temperature, 150 °C; source temperature, 80 °C.

Elemental analysis was conducted in the Microanalytical Laboratory, Department of Chemistry, School of Science, the University of Tokyo, using a vario MICRO cube elemental analyzer with addition of MgO.

CD spectra were recorded on a JASCO J-820 circular dichroism spectrometer. The experimental conditions were as follows: bandwidth, 1 nm; response, 0.5 s; data acquisition intervals, 0.5 nm; scanning rate, 100 nm/min; and number of scans, 4.

HPLC data were collected on a JASCO MD-4010 photodiode array detector connected with PU-4185-Binary RHPLC semi-micro-pump and CO-4060 column oven. The experimental conditions were as follows: slit width, 4 nm; data acquisition intervals, 4 nm; data integration width, 3 nm; and sampling rate, 5 points/s.

Single-crystal XRD analyses were performed using a Rigaku XtaLAB PRO MM007DW PILATUS diffractometer, and obtained data were processed by using CrysAlisPro 1.171.39.7e (Rigaku OD, 2015) software and analyzed by Olex [2] 1.2.10 (OlexSys Ltd., 2018) software [64], using SHELXL-2017/1 [65]. The crystals of the Zn^{II}**L2** complexes are handled in open air quickly before they decompose.

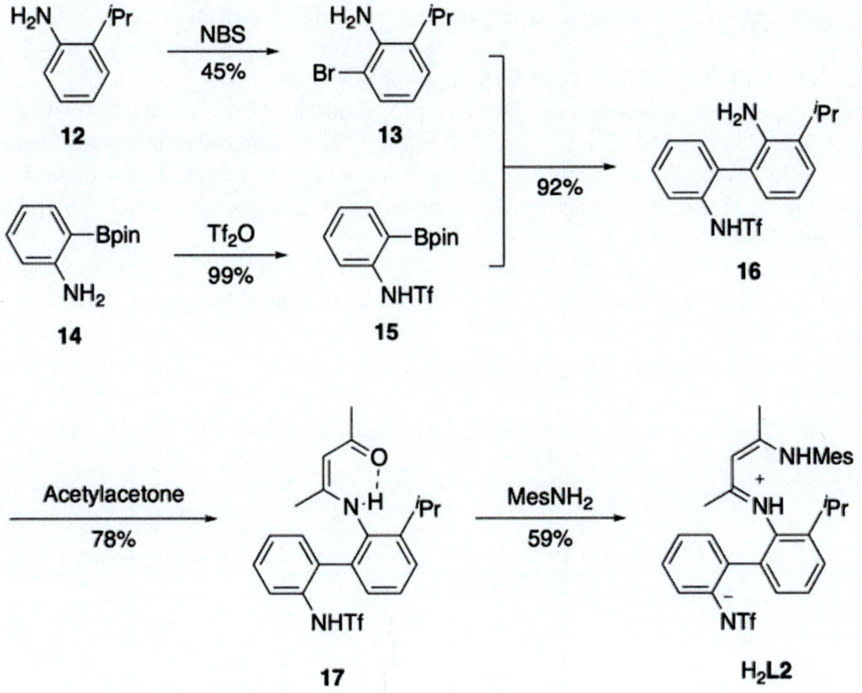

Fig. 3.23 Synthetic route to H_2**L2**

3.10.2 Synthesis of the Ligand

3.10.2.1 Synthetic Route to (3′-Isopropyl-2′-((4-(Mesitylamino)pent-3-en-2-Ylidene)Ammonio)-[1,1′-Biphenyl]-2-yl)((Trifluoromethyl)Sulfonyl)Amide (H_2L2)

H_2**L2** was synthesized via the route shown in Fig. 3.23.

3.10.2.2 2-Bromo-6-Isopropylaniline (13)

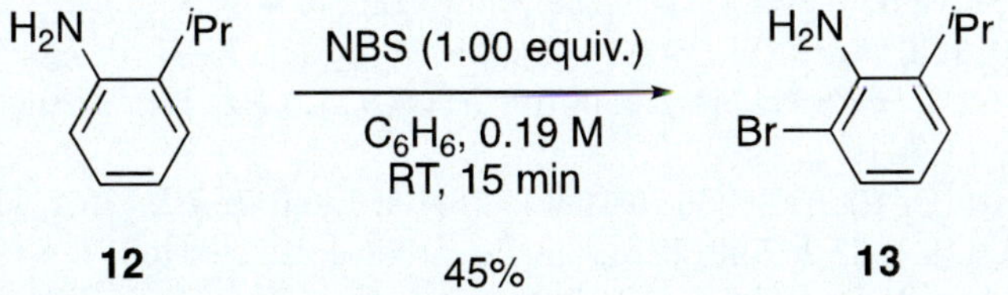

This compound was prepared in a procedure modified from a report in the literature [66].

NBS was recrystallized from water in prior to use.

A 100-mL flask was charged with 2-isopropylaniline (**12**) (1.40 mL, 10.0 mmol), benzene (54 mL), and NBS (1.78 g, 1.00 equiv.). The mixture was stirred for 15 min. The volatiles were removed under a reduced pressure. The crude product was purified by silica gel column chromatography using hexane/CH_2Cl_2 to give **13** as a colorless liquid (0.965 g, 45%).

^{1}H NMR ($CDCl_3$, 300 K, 500 MHz): δ 7.29 (dd, $J = 8.0, 1.3$ Hz, 1H), 7.08 (dd, $J = 7.7, 0.8$ Hz, 1H), 6.63 (t, $J = 7.8$ Hz, 1H), 4.15 (s, 2H), 2.90 (septet, $J = 6.8$ Hz, 1H), 1.26 (d, $J = 6.8$ Hz, 6H).

^{13}C NMR ($CDCl_3$, 300 K, 126 MHz): δ 141.0, 133.8, 129.9, 124.4, 119.2, 110.4, 28.8, 22.1.

HR-ESI-MS (positive, MeOH): *m/z* 214.0225 (required, 214.0226 for [**13**·H]$^+$ ($C_9H_{13}BrN^+$)).

3.10.2.3 1,1,1-Trifluoro-*N*-(2-(4,4,5,5-Tetramethyl-1,3,2-Dioxaborolan-2-yl)Phenyl)Methanesulfonamide (15)

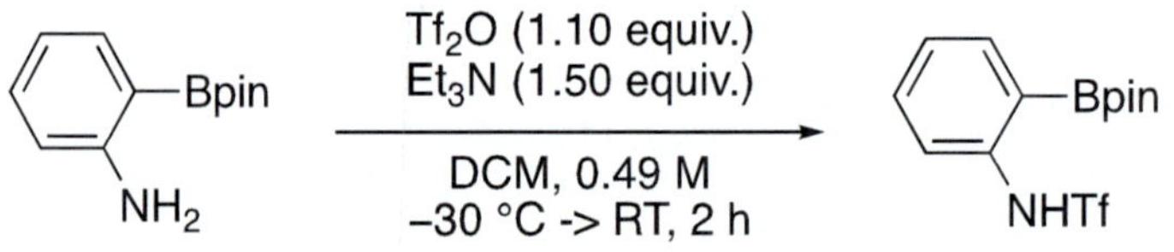

14 99% **15**

Et_3N was distilled and dehydrated over KOH in prior to use. Tf_2O was distilled over P_2O_5 in prior to use.

A three-necked 100-mL flask was charged with 2-(4,4,5,5-tetramethyl-1,3,2-dioxaborolan-2-yl)aniline (**14**) (4.05 g, 18.5 mmol), dehydrated CH_2Cl_2 (38 mL), and Et_3N (3.86 μL, 1.50 equiv.) under an Ar atmosphere. The mixture was cooled to −30 °C, and Tf_2O (3.34 μL, 1.10 equiv.) was added dropwise to the mixture. The mixture was stirred at −30 °C for 13 min and then at RT for 2 h. 1 M HClaq (38 mL) was added to the mixture, and the organic layer was separated. The aqueous layer was extracted with DCM (38 mL) twice. The extracts were washed with water (38 mL) and brine (38 mL) and dehydrated over Na_2SO_4. The volatiles were removed under a reduced pressure to give **15** as a pale yellow solid (6.45 g, 99%).

^{1}H NMR ($CDCl_3$, 300 K, 500 MHz): δ 9.10 (s, 1H), 7.79 (dd, $J = 7.4, 1.6$ Hz, 1H), 7.63 (d, $J = 8.3$ Hz, 1H), 7.48 (td, $J = 7.9, 1.6$ Hz, 1H), 7.20 (td, $J = 7.4, 0.9$ Hz, 1H), 1.38 (s, 12H).

^{13}C NMR ($CDCl_3$, 300 K, 126 MHz): δ 141.4, 136.6, 133.2, 125.1, 119.9 (q, $J^1_{CF} = 324$ Hz), 119.1, 85.2, 24.8. The aromatic carbon connected to B was not observed due to quadrupolar interaction with ^{10}B and ^{11}B.

^{19}F NMR ($CDCl_3$, 300 K, 471 MHz): δ −75.37.

^{11}B NMR ($CDCl_3$, 300 K, 160 MHz): δ −30.36.

HR-ESI-MS (negative, MeCN): *m*/*z* 349.0890 (required, 349.0887 for $[H_{-1}\mathbf{15}]^-$ ($C_{13}H_{16}BF_3NO_4S^-$)).

3.10.2.4 *N*-(2′-Amino-3′-Isopropyl-[1,1′-Biphenyl]-2-yl)-1,1,1-Trifluoromethanesulfonamide (16)

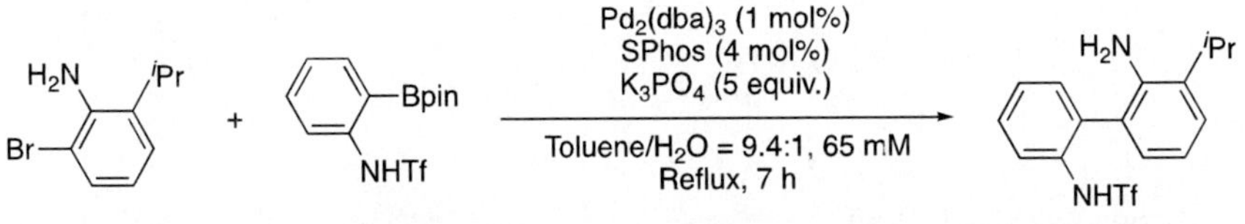

13 **15** (1.0 equiv.) **16**

A three-necked 200-mL flask was charged with **15** (2.13 g, 6.07 mmol), K_3PO_4 (6.27 g, 5 equiv.), H_2O (8.7 mL), toluene (82 mL), and **13** (1.25 g, 1.0 equiv.) under an Ar atmosphere. The mixture was bubbled with Ar. To the mixture were added $Pd_2(dba)_3$ (54.0 mg, 1 mol%) and SPhos (95.8 mg, 4 mol%). The mixture was heated at reflux for 7 h. The mixture was dehydrated with Na_2SO_4 and filtered through Celite. The volatiles were removed under a reduced pressure. To the residue were added $CHCl_3$ (44 mL) and 1 M HClaq (44 mL), and the organic layer was separated. The aqueous layer was extracted with $CHCl_3$ (44 mL) twice. The extracts were washed with water (44 mL) and brine (44 mL) and dehydrated over Na_2SO_4. The volatiles were removed under a reduced pressure. The crude product was purified by silica gel column chromatography using *n*-hexane/EtOAc to give **16** as an orange-yellow solid (2.00 g, 92%).

^{1}H NMR ($CDCl_3$, 300 K, 500 MHz): δ 8.81 (br, 1H), 7.60 (d, *J* = 7.7 Hz, 1H), 7.46–7.39 (m, 3H), 7.28 (dd, *J* = 7.6, 1.5 Hz, 1H), 7.03 (dd, *J* = 7.6, 1.6 Hz, 1H), 6.98 (t, *J* = 7.6 Hz, 1H), 3.99 (br, 2H), 2.98 (7, *J* = 6.8 Hz, 1H), 1.36 (d, *J* = 6.8 Hz, 3H), 1.29 (d, *J* = 6.8 Hz, 3H).

^{19}F NMR ($CDCl_3$, 300 K, 471 MHz): δ −77.12.

HR-ESI-MS (negative, MeCN): *m*/*z* 357.0900 (required, 357.0890 for $[H_{-1}\mathbf{16}]^-$ ($C_{16}H_{16}F_3N_2O_2S^-$)).

3.10.2.5 1,1,1-Trifluoro-*N*-(3′-Isopropyl-2′-((4-Oxopent-2-en-2-yl)Amino)-[1,1′-Biphenyl]-2-yl)Methanesulfonamide (17)

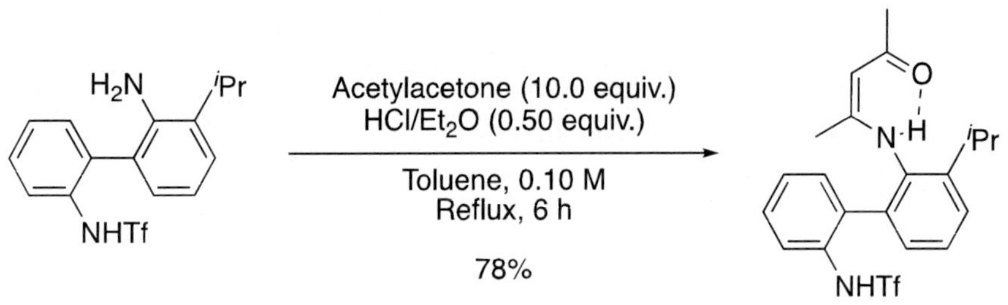

16 **17**

A three-necked 100-mL flask was charged with **16** (1.98 g, 5.53 mmol), dehydrated toluene (50 mL), acetylacetone (5.27 mL, 10.0 equiv.), a Et_2O solution of HCl (1.0 M, 2.56 mL, 0.50 equiv.) under an Ar atmosphere. The mixture was heated at reflux for 6 h. To the mixture was added saturated $NaHCO_3$aq (50 mL), and the organic layer was separated. The aqueous layer was extracted with toluene (50 mL) twice. The extracts were washed with water (50 mL) and brine (50 mL) and dehydrated over Na_2SO_4. The volatiles were removed under a reduced pressure. The crude product was purified by silica gel column chromatography twice, using *n*-hexane/EtOAc and CH_2Cl_2/EtOAc, respectively, to give **17** as a yellow solid (1.90 g, 78%).

In the NMR spectra, three isomers were observed in the ratio of 1:0.11:0.067. The second isomer would be a rotamer regarding the slow rotation around the biphenyl and phenyl–enaminone linkages. The last isomer would be an imine tautomer. Most signals of the minor isomers could not be located because of overlapping and low intensity. The major isomer showed broadening and changes in the chemical shifts at high concentrations.

^{1}H NMR ($CDCl_3$, 300 K, 500 MHz): δ (the major isomer) 7.44–7.22 (m, 7H), 4.86 (s, 1H), 3.22–3.11 (m, 1H), 1.99 (s, 3H), 1.38 (s, 3H), 1.26, (d, $J = 6.8$ Hz, 3H), 1.25 (d, $J = 6.8$ Hz, 3H); (the minor rotamer) 7.61 (d, $J = 7.7$ Hz, 1H), 4.94 (s, 1H), 1.47 (s, 3H); (the minor tautomer) 12.21 (s, 1H), 3.41 (AB pattern, $J = 15.5$ Hz, 2H), 2.89 (septet, $J = 6.8$ Hz, 1H), 2.12 (s, 3H), 1.17 (d, $J = 6.8$ Hz, 3H).

^{13}C NMR ($CDCl_3$, 300 K, 126 MHz): δ (the major isomer) 197.0, 163.4, 146.1, 137.6, 134.4, 132.05, 132.00, 130.8, 129.05, 129.01, 128.39, 128.25, 128.09, 126.2, 96.3, 28.55, 28.38, 23.8, 22.8, 19.5. Some signals could not be located due to broadening and overlap.

^{19}F NMR ($CDCl_3$, 300 K, 471 MHz): δ −76.77 (the major isomer), −76.03 (the minor rotamer), −77.30 (the minor tautomer).

HR-ESI-MS (negative, MeCN): *m*/*z* 439.1311 (required, 439.1309 for $[H_{-1}\mathbf{17}]^-$ ($C_{21}H_{22}F_3N_2O_3S^-$)).

3.10.2.6 Ligand H_2L2

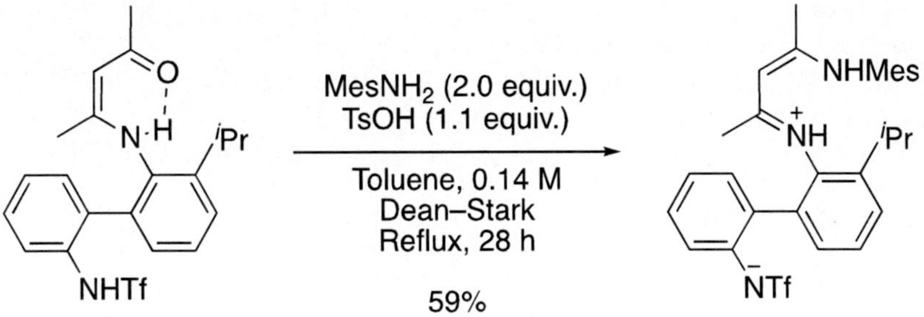

17 H_2**L2**

A 10-mL flask was charged with **17** (0.254 g, 0.577 mmol), $MesNH_2$ (165 μL, 2.0 equiv.), $TsOH{\cdot}H_2O$ (120 mg, 1.1 equiv.), dehydrated toluene (4.0 mL) under an Ar atmosphere, and equipped with a Dean–Stark apparatus filled with MS 4A and dehydrated toluene and with a reflux condenser. The mixture was heated at reflux for 28 h. To the mixture was added saturated $NaHCO_3$aq (5.0 mL), and the organic layer was separated. The aqueous layer was extracted with CH_2Cl_2 (5.0 mL) three times. The extracts were washed with water (5.0 mL) and brine (5.0 mL) and dehydrated over Na_2SO_4. The volatiles were removed under a reduced pressure. The crude product was recrystallized from CH_2Cl_2 by slow diffusion of *n*-hexane to give H_2**L2** as a colorless solid (0.190 g, 59%).

In the NMR spectra, three isomers were observed in a ratio of 1:0.20:0.11. These isomers were presumably rotamers arising from the slow rotation around the biphenyl and phenyl–diketimine linkages and the diketimine moiety. The NMR measurement at 343 K showed coalescence of the second and third isomers, and a different ratio of 1:0.26, supporting that these are interchangeable isomers. Several signals of the minor isomers could not be located because of overlapping and low intensity.

^{1}H NMR (CD_3CN, 300 K, 500 MHz): δ (the first isomer) 11.18 (br, 1H), 8.12 (br, 1H), 7.33 (dd, J = 8.1, 1.0 Hz, 1H), 7.25-7.20 (m, 3H), 7.08 (d, J = 8.0 Hz, 2H), 7.03 (dd, J = 7.6, 1.7 Hz, 1H), 6.96 (td, J = 7.4, 1.1 Hz, 1H), 6.80 (s, 1H), 6.75 (s, 1H), 4.21 (s, 1H), 2.75 (septet, J = 6.8 Hz, 1H), 2.36 (s, 3H), 2.29 (s, 3H), 2.23 (s, 3H), 1.97 (s, 3H), 1.73 (s, 3H), 1.04 (d, J = 6.8 Hz, 3H), 1.02 (d, J = 6.9 Hz, 3H); (the second isomer) 12.02 (br, 1H), 8.22 (br, 1H), 7.49-7.37 (m, 4H), 7.02-6.98 (m, J = 6.6 Hz, 2H), 4.50 (s, 1H), 3.05 (septet, J = 6.9 Hz, 1H), 2.84 (s, 3H), 2.27 (s, 3H), 2.09 (s, 3H), 1.99 (s, 3H), 1.58 (s, 3H), 1.33 (d, J = 6.9 Hz, 3H), 1.18 (d, J = 6.8 Hz, 3H); (the third isomer) 8.43 (br, 1H), 5.23 (br, 1H), 3.10 (septet, J = 6.8 Hz, 1H), 2.25 (s, 3H), 2.03 (s, 3H), 1.35 (d, J = 6.8 Hz, 3H).

^{13}C NMR (CD_3CN, 300 K, 126 MHz): δ (the major isomer) 170.9, 169.9, 145.1, 144.5, 139.7, 139.4, 135.8, 135.3, 134.6, 133.8, 132.0, 131.8, 130.06, 129.95, 129.84, 129.18, 129.17, 125.7, 125.1, 122.8 (q, J^1_{CF} = 328 Hz), 122.3, 91.1, 28.9, 25.3, 22.5, 21.9, 21.4, 21.0, 17.84, 17.67; (the minor isomers) 172.8, 168.9, 145.19, 145.10,

144.5, 144.1, 140.1, 139.77, 139.66, 139.4, 136.02, 135.99, 132.3, 131.5, 130.32, 130.27, 130.16, 129.5, 129.3, 127.3, 125.8, 123.0, 93.0, 29.2, 25.4, 21.84, 21.78, 21.1, 17.90, 17.80.

^{19}F NMR (CD_3CN, 300 K, 471 MHz): δ (the first isomer) −78.49; (the second isomer) −78.87; and (the third isomer) −78.52.

HR-ESI-MS (negative, MeCN): *m/z* 556.2256 (required, 556.2251 for $[H_{-1}\mathbf{L2}]^-$ ($C_{30}H_{33}F_3N_3O_2S^-$)).

Elemental analysis (calcd. for $C_{30}H_{34}F_3N_3O_2S$ ($H_2\mathbf{L2}$), found): C (64.61, 64.78), H (6.15, 6.21), N (7.54, 7.54).

m.p.: 242.5–243.1 °C

3.10.3 Syntheses of the Metal Complexes

3.10.3.1 Complex (*R**,*R**)-[$Zn_2\mathbf{L2}_2$]

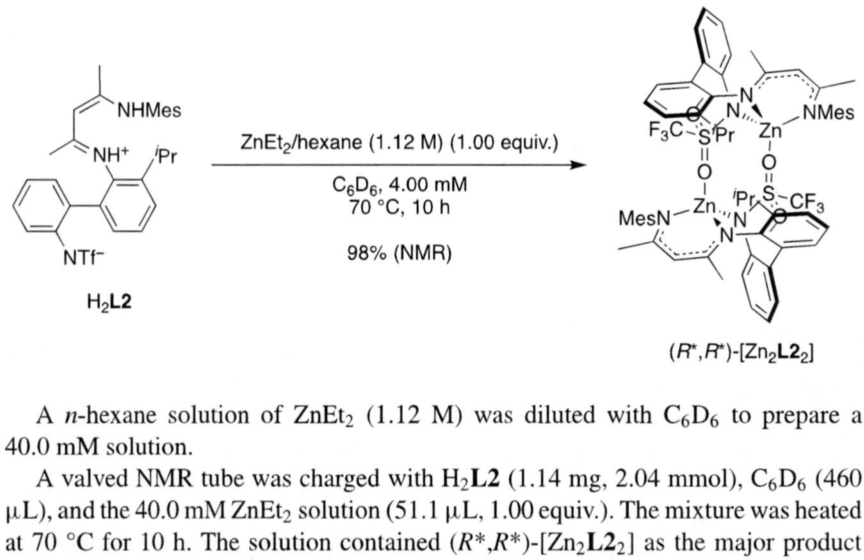

A *n*-hexane solution of $ZnEt_2$ (1.12 M) was diluted with C_6D_6 to prepare a 40.0 mM solution.

A valved NMR tube was charged with $H_2\mathbf{L2}$ (1.14 mg, 2.04 mmol), C_6D_6 (460 μL), and the 40.0 mM $ZnEt_2$ solution (51.1 μL, 1.00 equiv.). The mixture was heated at 70 °C for 10 h. The solution contained (*R**,*R**)-[$Zn_2\mathbf{L2}_2$] as the major product (98% as determined by ^{1}H NMR using an internal standard).

Broadening of signals was observed in NMR spectra, suggesting a dynamic exchange with an *R**,*S** diastereomer.

^{1}H NMR (C_6D_6, 338 K, 500 MHz): δ 7.42 (d, J = 7.5 Hz, 1H), 7.20–7.18 (m, 2H), 7.08–7.06 (m, 3H), 7.00–6.97 (m, 2H), 6.88 (s, 1H), 4.49 (s, 1H), 3.13 (septet, J = 6.8 Hz, 1H), 2.28 (s, 3H), 2.15 (s, 3H), 2.09 (s, 3H), 1.50 (s, 3H), 1.39 (s, 3H), 1.16 (d, J = 6.8 Hz, 3H), 1.16 (d, J = 6.8 Hz, 3H).

^{19}F NMR (C_6D_6, 338 K, 471 MHz): δ −75.76.

3.10.3.2 Complex (*S*)-[ZnL2((*S*)-dpp)] ((*S*)-dpp = (*S*)-α,α-Diphenyl-2-Pyrrolidinemethanol)

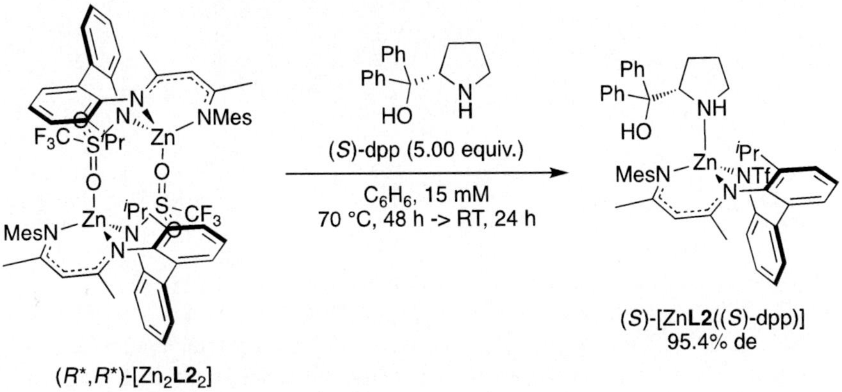

A 50-mL Schlenk tube was charged with H_2**L2** (87.2 mg, 0.156 mmol), C_6H_6 (4.67 mL), and a toluene solution of $ZnEt_2$ (1.12 M, 154 μL, 1.10 equiv.). The mixture was heated at 70 °C for 19 h. To the mixture was added a C_6H_6 solution of (*S*)-dpp (1.00 M, 391 μL, 2.50 equiv.). The mixture was stirred at 70 °C for 48 h and then at room temperature for 24 h. The volatiles were removed under a reduced pressure to give a colorless liquid containing (*S*)-[Zn**L2**((*S*)-dpp)] (95.4% de as determined by ^{1}H NMR) and (*S*)-dpp. The crude mixture was used for the next reaction without further purification.

^{1}H NMR (C_6D_6, 300 K, 500 MHz): δ ((*S*)-[Zn**L2**((*S*)-dpp)]) 8.16 (d, *J* = 7.7 Hz, 1H), 7.35 (dd, *J* = 5.6, 3.4 Hz, 1H), 7.31 (d, *J* = 6.8 Hz, 2H), 7.22 (td, *J* = 7.7, 1.4 Hz, 1H), 7.19–7.00 (m, 10H), 6.92 (t, *J* = 7.7 Hz, 2H), 6.86 (t, *J* = 7.3 Hz, 1H), 6.64 (s, 1H), 5.89 (t, *J* = 6.1 Hz, 1H), 5.62 (s, 1H), 4.67 (s, 1H), 4.33 (s, 1H), 3.77 (dd, *J* = 8.3, 1.8 Hz, 1H), 3.24 (septet, *J* = 6.8 Hz, 1H), 2.86–2.79 (m, 4H), 2.26–2.19 (m, 1H), 1.95–1.87 (m, 4H), 1.83–1.78 (m, 4H), 1.44 (s, 3H), 1.26 (s, 3H), 1.19 (d, *J* = 7.0 Hz, 3H), 1.15 (d, *J* = 6.7 Hz, 3H), 0.81 (dt, *J* = 13.3, 6.9 Hz, 1H), 0.48 (s, 1H), −0.38– − 0.48 (m, 1H); ((*S*)-dpp) 7.67 (br, 2H), 7.60 (d, *J* = 7.5 Hz, 2H), 7.19-7.16 (m, 4H), 7.15-7.00 (m, 2H), 4.67 (br, 1H), 3.84 (br, 1H), 2.62 (br, 1H), 2.39 (br, 1H), 1.67 (br, 1H), 1.48–1.44 (m, 1H), 1.32–1.31 (m, 2H), 1.17–1.04 (m, 1H).

^{19}F NMR (C_6D_6, 300 K, 471 MHz): δ ((*S*)-[Zn**L2**((*S*)-dpp)]) −73.99; ((*R*)-[Zn**L2**((*S*)-dpp)]) −74.70; ((*R*)-[Zn(H**L2**)((*S*)-H_{-1}dpp)]) −77.64.

3.10.3.3 Complex (*S*)-[ZnL2(NC^{t}Bu)]

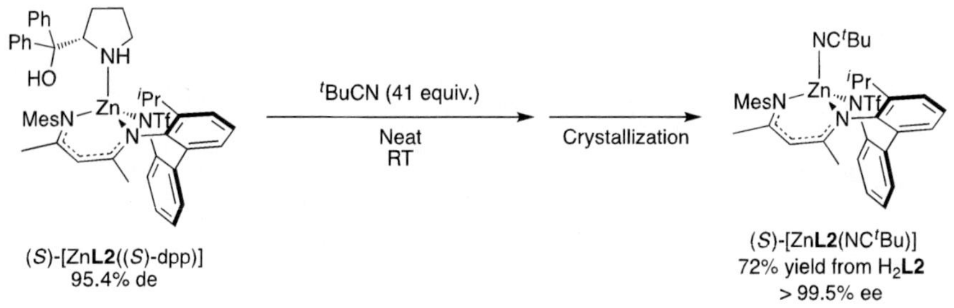

A 40-mL vial was charged with the aforementioned mixture of (*S*)-[Zn**L2**((*S*)-dpp)] and (*S*)-dpp (from 0.156 mmol of H_2**L2**), ^{t}BuCN (0.70 mL, 41 equiv.), C_6H_6 (2.6 mL), and HMDSO (36 mL). The mixture was left at room temperature for 7 days. The formed crystals were collected by decantation and washed with HMDSO (20 mL) twice to give (*S*)-[Zn**L2**(NC^{t}Bu)] as a colorless solid (78.7 mg, 72% from H_2**L2**). The enantiopurity was determined to be >99.5% ee by ^{19}F NMR using (*R*)-methyl *p*-tolyl sulfoxide ((*R*)-mts) (5 equiv.) as a chiral shift reagent and assuming the lower detection limit as $S/N = 3$.

^{1}H NMR (CD_2Cl_2, 300 K, 500 MHz): δ 7.36 (dd, $J = 8.0, 0.8$ Hz, 1H), 7.34 (dd, $J = 7.8, 1.3$ Hz, 1H), 7.29 (td, $J = 7.6, 1.6$ Hz, 1H), 7.22–7.18 (m, 2H), 7.14 (dd, $J = 7.5, 1.6$ Hz, 1H), 7.07 (d, $J = 7.4$ Hz, 1H), 6.98 (s, 1H), 6.88 (s, 1H), 4.45 (s, 1H), 3.13 (septet, $J = 6.9$ Hz, 1H), 2.28 (s, 3H), 2.24 (s, 3H), 2.11 (s, 3H), 1.55 (s, 3H), 1.52 (s, 3H), 1.28 (d, $J = 6.9$ Hz, 3H), 1.24 (d, $J = 6.9$ Hz, 3H), 1.15 (s, 9H).

^{13}C NMR (CD_2Cl_2, 300 K, 126 MHz): δ 168.5, 167.8, 144.07, 143.93, 141.5, 140.0, 139.1, 136.3, 134.2, 132.3, 131.55, 131.44, 129.72, 129.53, 129.45, 128.9, 127.8, 127.3, 125.75, 125.61, 125.2, 121.2 (q, $J^1_{CF} = 325$ Hz), 92.6, 28.7, 27.8, 27.0, 24.5, 23.6, 23.15, 23.01, 20.9, 18.32, 18.27.

^{19}F NMR (CD_2Cl_2, 300 K, 471 MHz): δ −78.49.

^{19}F NMR (C_6D_6, 300 K, 471 MHz) with (*R*)-mts (5 equiv.): δ −76.06 (S/N = 1282).

Elemental analysis (calcd. for $C_{36}H_{41}F_3N_4O_2SZn$ ([Zn**L2**(NC^{t}Bu)]), found): C (59.70, 59.80), H (5.87, 5.87), N (7.96, 7.90).

3.10.3.4 Complex (*R*)-[ZnL2(NC^{t}Bu)]

This compound was synthesized similarly to (*S*)-[Zn**L2**(NC^{t}Bu)], using (*R*)-dpp instead of (*S*)-dpp. The ^{1}H and ^{19}F NMR data matched those of (*S*)-[Zn**L2**(NC^{t}Bu)]. The ee was > 99.5% as determined by ^{19}F NMR using (*R*)-mts (5 equiv.) as a chiral shift reagent and assuming the lower detection limit as $S/N = 3$.

^{19}F NMR (C_6D_6, 300 K, 471 MHz) with (*R*)-mts (5 equiv.): δ −76.27 (S/N = 1521).

3.10.3.5 Complex [ZnL2(9)] (9 = 1-Naphthaldehyde)

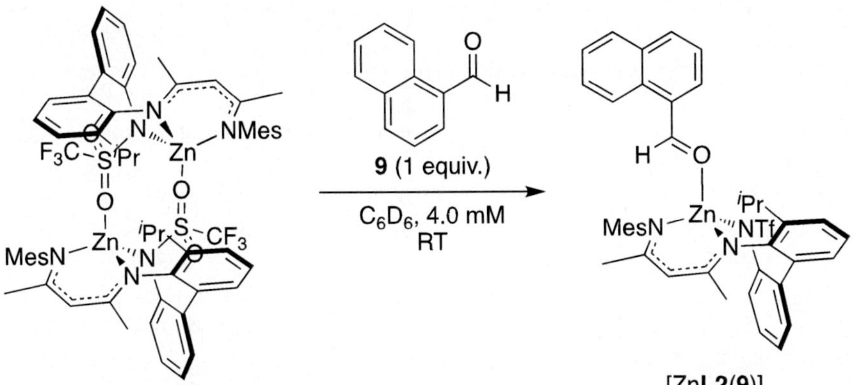

(R*,R*)-[Zn$_2$**L2**$_2$]

[Zn**L2**(**9**)]
(racemic)

Compound **9** was distilled and dehydrated over MS 4A in prior to use. A *n*-hexane solution of $ZnEt_2$ (1.12 M) was diluted with C_6D_6 to prepare a 40.0 mM solution.

A valved NMR tube was charged with H_2**L2** (1.20 mg, 2.15 μmol), C_6D_6 (484 μL), and the 40.0 mM $ZnEt_2$ solution (59.3 μL, 1.00 equiv.). The mixture was heated at 70 °C for 22 h. To the mixture was added a C_6D_6 solution of **9** (199 mM, 10.8 μL, 1.00 equiv.). The solution contained [Zn**L2**(**9**)] as the major product.

^{1}H NMR (C_6D_6, 300 K, 500 MHz): δ 9.29 (br, 1H), 8.73 (br, 1H), 8.08 (d, *J* = 8.0 Hz, 1H), 7.46–7.43 (m, 2H), 7.37–7.31 (m, 4H), 7.23–7.16 (m, 4H), 7.06 (td, *J* = 7.4, 0.9 Hz, 1H), 6.92 (s, 1H), 6.86 (t, *J* = 7.6 Hz, 1H), 6.30 (s, 1H), 4.46 (s, 1H), 3.48 (septet, *J* = 6.8 Hz, 1H), 2.62 (s, 3H), 2.08 (s, 3H), 1.78 (s, 3H), 1.64 (s, 3H), 1.45 (d, *J* = 6.8 Hz, 3H), 1.41 (s, 3H), 1.29–1.21 (m, 3H). The chemical shifts of the signals arising from **9** varied by a slight error in the amount of **9**, indicating fast exchange of coordinated and excess **9**.

^{19}F NMR (C_6D_6, 300 K, 471 MHz): δ −76.98.

3.10.4 Single-Crystal X-Ray Diffraction Analyses

3.10.4.1 Complex (R*,R*)-[Zn$_2$L2$_2$]

A single crystal suitable for measurement was grown by liquid–liquid diffusion of HMDSO into a C_6D_6 solution.

Crystal data for (*R**,*R**)-[Zn$_2$**L2**$_2$] ($C_{60}H_{64}F_6N_6O_4S_2Zn_2$) ($M$ = 1244.04 g/mol): triclinic, space group *P* − 1 (no. 2), a = 12.40390(10) Å, b = 13.19290(10) Å, c = 20.56270(10) Å, α = 103.9800(10)°, β = 97.9590(10)°, γ = 111.7010(10)°, V = 2935.43(4) Å^3, Z = 2, T = 93.15 K, μ(CuKα) = 2.262 mm^{-1}, D_{calc} = 1.407 g/cm^3,

249,205 reflections measured (4.576° ≤ 2Θ ≤ 147.618°), 11,724 unique (R_{int} = 0.0555, R_{sigma} = 0.0162) which were used in all calculations. The final R_1 was 0.0318 ($I > 2\sigma(I)$), and wR_2 was 0.0836 (all data).

3.10.4.2 Complex (*S*)-[ZnL2((*S*)-dpp)]

A single crystal suitable for measurement was grown by liquid–liquid diffusion of HMDSO into a C_6H_6 solution.

Crystal data for (*S*)-[Zn**L2**((*S*)-dpp)]·1/2C_6H_6 ($C_{50}H_{54}F_3N_4O_3SZn$) (M = 913.40 g/mol): orthorhombic, space group $P2_12_12_1$ (no. 19), a = 9.69090(10) Å, b = 28.5457(2) Å, c = 33.1549(3) Å, V = 9171.76(14) Å^3, Z = 8, T = 93.15 K, μ(CuKα) = 1.648 mm^{-1}, D_{calc} = 1.323 g/cm^3, 42,704 reflections measured (5.33° ≤ 2Θ ≤ 130.174°), 15,631 unique (R_{int} = 0.0330, R_{sigma} = 0.0399) which were used in all calculations. The final R_1 was 0.0266 ($I > 2\sigma(I)$), and wR_2 was 0.0662 (all data).

3.10.4.3 Complex (*S*)-[ZnL2(NC^{t}Bu)]

A single crystal suitable for measurement was grown by liquid–liquid diffusion of HMDSO into a solution in ^{t}BuCN/C_6H_6 = 1:3.

Crystal data for (*S*)-[Zn**L2**(NC^{t}Bu)] ($C_{35}H_{41}F_3N_4O_2SZn$) (M = 704.15 g/mol): orthorhombic, space group $P2_12_12_1$ (no. 19), a = 9.25690(10) Å, b = 17.53060(10) Å, c = 21.7964(2) Å, V = 3537.10(5) Å^3, Z = 4, T = 93.15 K, μ(CuKα) = 1.948 mm^{-1}, D_{calc} = 1.322 g/cm^3, 16,973 reflections measured (6.47° ≤ 2Θ ≤ 146.968°), 6944 unique (R_{int} = 0.0183, R_{sigma} = 0.0195) which were used in all calculations. The final R_1 was 0.0223 ($I > 2\sigma(I)$), and wR_2 was 0.0613 (all data). The final Flack parameter was −0.002(5).

3.10.4.4 Complex [ZnL2(9)]

A single crystal suitable for measurement was grown by liquid–liquid diffusion of *n*-hexane into a C_6D_6 solution.

Crystal data for [Zn**L2**(**9**)] ($C_{41}H_{40}F_3N_3O_3SZn$) (M = 777.19 g/mol): monoclinic, space group $C2/c$ (no. 15), a = 21.83508(12) Å, b = 9.4013(5) Å, c = 36.2443(2) Å, β = 96.9370(5)°, V = 7385.7(4) Å^3, Z = 8, T = 93.15 K, μ(CuKα) = 1.939 mm^{-1}, D_{calc} = 1.398 g/cm^3, 80,173 reflections measured (4.912° ≤ 2Θ ≤ 146.958°), 7361 unique (R_{int} = 0.0391, R_{sigma} = 0.0151) which were used in all calculations. The final R_1 was 0.0294 ($I > 2\sigma(I)$), and wR_2 was 0.0769 (all data).

3.10.5 *Miscellaneous Experiments*

A *n*-hexane solution of $ZnEt_2$ (1.12 M) was diluted with C_6D_6 to prepare a 40.0 mM solution.

3.10.5.1 Time-Course Study of Asymmetric Induction

MS 4A was dried under vacuum at approximately 450 °C for 5 min.

A valved NMR tube was charged with H_2**L2** (1.63 mg, 2.92 μmol), dry MS 4A (~10 mg), C_6D_6 (658 μL), and the 40.0 mM solution of $ZnEt_2$ (73.1 μL, 1.00 equiv.). The mixture was heated at 70 °C for 23 h. To the mixture was added a C_6D_6 solution of (*S*)-dpp (100 mM, 73.1 μL, 2.50 equiv.). The mixture was heated at 70 °C for 48 h.

3.10.5.2 Control Experiment with (*S*)-2-(Methoxydiphenylmethyl) Pyrrolidine ((*S*)-mdp)

A valved NMR tube was charged with H_2**L2** (1.17 mg, 2.10 μmol), dry MS 4A (~10 mg), C_6D_6 (472 μL), and the 40.0 mM solution of $ZnEt_2$ (52.5 μL, 1.00 equiv.). The mixture was heated at 70 °C for 23 h. To the mixture was added a C_6D_6 solution of (*S*)-mdp (100 mM, 52.5 μL, 2.50 equiv.). The mixture was heated at 70 °C for 72 h.

3.10.5.3 Asymmetric Induction with 1 Equiv. of (*S*)-dpp

A valved NMR tube was charged with H_2**L2** (1.15 mg, 2.06 μmol), dry MS 4A (~10 mg), C_6D_6 (464 μL), and the 40.0 mM solution of $ZnEt_2$ (51.6 μL, 1.00 equiv.). The mixture was heated at 70 °C for 23 h. To the mixture was added a C_6D_6 solution of (*S*)-dpp (100 mM, 20.6 μL, 1.00 equiv.). The mixture was heated at 70 °C for 48 h.

3.10.5.4 Test of (*R*)-mts as a Chiral Shift Reagent

A valved NMR tube was charged with H_2**L2** (1.12 mg, 2.01 μmol), C_6D_6 (447 μL), and the 40.0 mM solution of $ZnEt_2$ (55.2 μL, 1.10 equiv.). The mixture was heated at 70 °C for 24 h. To the mixture was added a C_6D_6 solution of (*R*)-mts (88 mM, 25.1 μL, 1.10 equiv.).

3.10.5.5 Recovery of (*S*)-dpp After the Use

The synthesis of (*S*)-[Zn**L2**(NC^{t}Bu)] was conducted by the aforementioned procedure using 24.4 mg of (*S*)-dpp. After the crystals of (*S*)-[Zn**L2**(NC^{t}Bu)] were collected, the mother liquor and washings were combined. The volatiles were removed under a reduced pressure. The residue was purified by aminopropyl-modified silica gel column chromatography using *n*-hexane/EtOAc to give (*S*)-dpp as a colorless liquid (23.3 mg, 95%).

3.10.5.6 Test of Configurational Stability of (*S*)-[ZnL2(NC^{t}Bu)]

A valved NMR tube was charged with (*S*)-[Zn**L2**(NC^{t}Bu)] (1.5 mg, 2.1 μmol) and C_6D_6 (533 μL). The mixture was left at 24 °C for 24 h. To the mixture was added a C_6D_6 solution of (*R*)-mts (800 mM, 13.3 μL, 5.0 equiv.).

A valved NMR tube was charged with (*S*)-[Zn**L2**(NC^{t}Bu)] (1.3 mg, 1.8 μmol) and C_6D_6 (462 μL). The mixture was heated at 70 °C for 24 h. To the mixture was added a C_6D_6 solution of (*R*)-mts (800 mM, 11.5 μL, 5.0 equiv.).

3.10.5.7 Enantioselective Catalysis

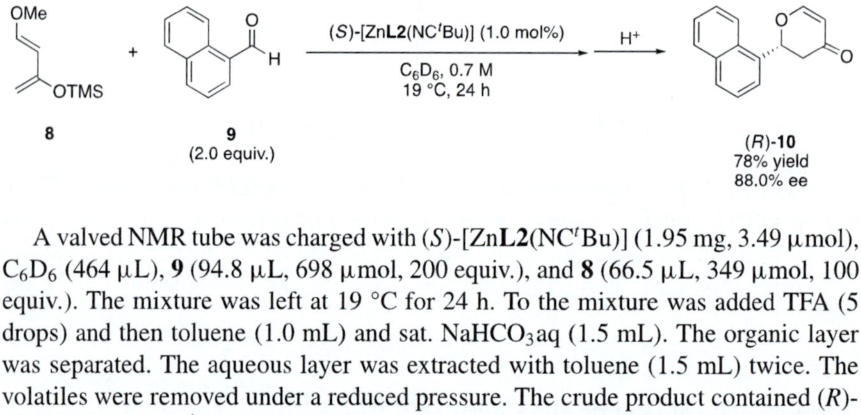

A valved NMR tube was charged with (*S*)-[Zn**L2**(NC^{t}Bu)] (1.95 mg, 3.49 μmol), C_6D_6 (464 μL), **9** (94.8 μL, 698 μmol, 200 equiv.), and **8** (66.5 μL, 349 μmol, 100 equiv.). The mixture was left at 19 °C for 24 h. To the mixture was added TFA (5 drops) and then toluene (1.0 mL) and sat. $NaHCO_3$aq (1.5 mL). The organic layer was separated. The aqueous layer was extracted with toluene (1.5 mL) twice. The volatiles were removed under a reduced pressure. The crude product contained (*R*)-**10** as judged from ^{1}H NMR in comparison with the reported data [62]. The yield was estimated to be 78% by ^{1}H NMR using an internal standard. The ee was determined to be 88.0% by HPLC after pretreatment with a short path of silica gel using EtOAc.

^{1}H NMR ($CDCl_3$, 300 K, 500 MHz): δ 8.00 (dd, J = 7.1, 1.2 Hz, 2H), 7.93 (d, J = 7.9 Hz, 2H), 6.19 (dd, J = 14.2, 3.4 Hz, 1H), 5.62 (dd, J = 6.0, 1.1 Hz, 1H), 3.09 (dd, J = 17.0, 14.2 Hz, 1H), 2.87 (ddd, J = 17.0, 3.4, 1.2 Hz, 1H). Some signals could not be located due to overlap with other species.

HPLC (Chiralcel OD, *n*-hexane/^{i}PrOH = 9:1, 1.0 mL/min, 25 °C): t_R 32.4 min (minor, *S*), 38.5 min (major, *R*).

The racemic sample for HPLC analysis was obtained by using (R^*,R^*)-[Zn_2**L2**$_2$] instead of (S)-[Zn**L2**(NC^{t}Bu)]. HPLC analysis showed 0.09% ee for **10**.

The absolute configuration of the product was determined by the analogy to the case of PhCHO. The reaction was conducted in a similar procedure using PhCHO instead of **9**, and the product was analyzed by HPLC (Chiralcel OD, n-hexane/^{i}PrOH = 9:1, 1.0 mL/min, 25 °C) to give t_R = 12.3 min (minor, S), 14.9 min (major, R). The absolute configurations of these products were determined in comparison with the reported data [63].

References

1. Blaser H-U (2013) Rendiconti Lincei 24:213–216
2. Jeschke P (2018) Pest Manag Sci 74:2389–2404
3. Abate A, Brenna E, Fuganti C, Gatti FG, Giovenzana T, Malpezzi L, Serra S (2005) J Org Chem 70:1281–1290
4. Sanchez-Carnerero EM, Agarrabeitia AR, Moreno F, Maroto BL, Muller G, Ortiz MJ, de la Moya S (2015) Chem Eur J 21:13488–13500
5. Phillips KES, Katz TJ, Jockusch S, Lovinger AJ, Turro NJ (2001) J Am Chem Soc 123:11899–11907
6. Jedrzejewska H, Szumna A (2017) Chem Rev 117:4863–4899
7. Shaver MP, Cameron DJA (2010) Biomacromol 11:3673–3679
8. Barron LD (2007) Space Sci Rev 135:187–201
9. Fujii N, Takata T, Fujii N, Aki K, Sakaue H (2018) Biochim Biophys Acta Proteins Proteomics 1866:840–847
10. Werner A (1911) Ber Dtsch Chem Ges 44:1887–1898
11. Knowles WS, Sabacky MJ (1968) Chem Commun 1445–1446
12. Mills WH, Gotts RA (1926) J Chem Soc 129:3121–3131
13. Deuschelcornioley C, Stoecklievans H, Vonzelewsky A (1990) J Chem Soc Chem Commun 121–122
14. Lennartson A, Håkansson M (2015) New J Chem 39:5936–5943
15. Ganter C (2003) Chem Soc Rev 32:130–138
16. Metcalf DH, Snyder SW, Demas JN, Richardson FS (1990) J Am Chem Soc 112:469–479
17. Desreux JF (1980) Inorg Chem 19:1319–1324
18. Brunner H (1969) Angew Chem Int Ed Engl 8:382–383
19. Pierre J-L (1998) Coord Chem Rev 178–180:1183–1192
20. von Zelewsky A (1999) Coord Chem Rev 190–192:811–825
21. Knof U, von Zelewsky A (1999) Angew Chem Int Ed 38:302–322
22. von Zelewsky A, Mamula O (2000) J Chem Soc Dalton Trans 219–231
23. Crassous J (2012) Chem Commun 48:9687–9695
24. Lacour J (2002) CHIMIA Int J Chem 56:672–675
25. Lacour J, Hebbe-Viton V (2003) Chem Soc Rev 32:373–382
26. Lacour J, Frantz R (2005) Org Biomol Chem 3:15–19
27. Lacour J, Moraleda D (2009) Chem Commun 7073–7089
28. Brunner H (2001) Eur J Inorg Chem 905–912
29. Smirnoff AP (1920) Helv Chim Acta 3:177–195
30. Cao Z-Y, Brittain WDG, Fossey JS, Zhou F (2015) Cat Sci Technol 5:3441–3451
31. Zhang L, Meggers E (2017) Chem Asian J 12:2335–2342
32. Meggers E (2017) Angew Chem Int Ed 56:5668–5675
33. Coughlin FJ, Westrol MS, Oyler KD, Byrne N, Kraml C, Zysman-Colman E, Lowry MS, Bernhard S (2008) Inorg Chem 47:2039–2048

34. Ovanesyan NS, Makhaev VD, Aldoshin SM, Gredin P, Boubekeur K, Train C, Gruselle M (2005) Dalton Trans 3101–3107
35. Wu K, Li K, Hou YJ, Pan M, Zhang LY, Chen L, Su CY (2016) Nat Commun 7:10487
36. Huang K, Ma Q, Shen X, Gong L, Meggers E, Asian J (2016) Org Chem 5:1198–1203
37. Eaton DR, Phillips WD, Caldwell DJ (1963) J Am Chem Soc 85:397–406
38. Brunner H, Miehling W (1983) Angew Chem Int Ed Engl 22:164–165
39. Brunner H, Faustmann P, Dietl A, Nuber B (1997) J Organomet Chem 542:255–263
40. Feng L, Geisselbrecht Y, Blanck S, Wilbuer A, Atilla-Gokcumen GE, Filippakopoulos P, Kraling K, Celik MA, Harms K, Maksimoska J, Marmorstein R, Frenking G, Knapp S, Essen LO, Meggers E (2011) J Am Chem Soc 133:5976–5986
41. Brunner H, Faustmann P, Nuber B (1998) J Organomet Chem 556:129–140
42. Wood TE, Dalgleish ND, Power ED, Thompson A, Chen X, Okamoto Y (2005) J Am Chem Soc 127:5740–5741
43. Wood TE, Ross AC, Dalgleish ND, Power ED, Thompson A, Chen X, Okamoto Y (2005) J Org Chem 70:9967–9974
44. Ali AA, Benson RE, Blumentritt S, Cameron TS, Linden A, Wolstenholme D, Thompson A (2007) J Org Chem 72:4947–4952
45. Hashimoto T, Nishimura T, Lim JM, Kim D, Maeda H (2010) Chem Eur J 16:11653–11661
46. Maeda H, Nishimura T, Akuta R, Takaishi K, Uchiyama M, Muranaka A (2013) Chem Sci 4:1204–1211
47. Maeda H, Nishimura T, Tsujii A, Takaishi K, Uchiyama M, Muranaka A (2014) Chem Lett 43:1078–1080
48. Furusho Y, Goto H, Itomi K, Katagiri H, Miyagawa T, Yashima E (2011) Chem Commun 47:9795–9797
49. Imamura T, Maehara T, Sekiya R, Haino T (2016) Chem Eur J 22:3250–3254
50. Harada K, Sekiya R, Maehara T, Haino T (2019) Org Biomol Chem 17:4729–4735
51. Meggers E (2010) Chem Eur J 16:752–758
52. Vallee BL, Auld DS (1990) Biochemistry 29:5647–5659
53. Prelog V (1964) Pure Appl Chem 9:119–130
54. Yadav CL, Rajput G, Bisht KK, Drew MGB, Singh N (2019) Inorg Chem 58:14449–14456
55. Becker JJ, White PS, Gagné MR (2001) J Am Chem Soc 123:9478–9479
56. Yang L, Powell DR, Houser RP (2007) Dalton Trans 955–964
57. Chamayou AC, Ludeke S, Brecht V, Freedman TB, Nafie LA, Janiak C (2011) Inorg Chem 50:11363–11374
58. Enamullah M, Makhloufi G, Ahmed R, Joy BA, Islam MA, Padula D, Hunter H, Pescitelli G, Janiak C (2016) Inorg Chem 55:6449–6464
59. Pellissier H (2009) Tetrahedron 65:2839–2877
60. Gutmann V (1976) Coord Chem Rev 18:225–255
61. Beckett MA, Brassington DS, Coles SJ, Hursthouse MB (2000) Inorg Chem Commun 3:530–533
62. White JD, Shaw S (2011) Org Lett 13:2488–2491
63. Du H, Long J, Hu J, Li X, Ding K (2002) Org Lett 4:4349–4352
64. Dolomanov OV, Bourhis LJ, Gildea RJ, Howard JAK, Puschmann HJ (2009) Appl Cryst 42:339–341
65. Sheldrick GM (2015) Acta Cryst C 71:3–8
66. Chupp JP (1980) US4188342A, 12 Feb

Chapter 4
Conclusions

4.1 Conclusions

Kinetically controlled stepwise synthesis is a useful method for preparation of metal complexes with intricate structures. However, its applicable scope has been limited chiefly to the inert metal complexes. In this thesis, I presented two new strategies to expand the target scope of kinetically controlled stepwise synthesis toward the labile metal complexes. These strategies belong to two distinct fields: a heterometallic complex and a chiral-at-metal complex (Fig. 4.1).

Heterometallic complexes show interesting properties based on interactions between different metal ions. However, it is difficult to selectively synthesize a heterometallic complex which contains analogous metal ions in desired numbers and positions. In Chap. 2, I proposed "site-selective redox switching and transmetalation" as a new stepwise method to selectively arrange different metals. This method consists of four steps: homometallic complexation, site-selective oxidation, site-selective transmetalation, and reduction. Using this method, selective synthesis of a Co^{II}–Ni^{II} heterometallic complex was achieved, which is normally difficult because of the lability and similarity of Co^{II} and Ni^{II} complexes. This result gives wider access to novel heterometallic complexes, which can exhibit functions specific to the combination of metals.

Chiral-at-metal complexes are useful as enantioselective catalysts or chiroptical materials because of coupling of their chirality and metal-based functions. However, tetrahedral chiral-at-metal complexes without chiral ligands are difficult to synthesize enantioselectively because of their fast stereoinversion. In Chap. 3, I proposed an unsymmetrical tridentate ligand with strong and non-planar coordination ability as a ligand design to kinetically stabilize a metal complex against stereoinversion. Using this ligand in combination with an acidic chiral auxiliary ligand, enantioselective synthesis of a tetrahedral chiral-at-Zn^{II} complex was achieved, which is normally difficult because of the lability of Zn^{II} complexes and tetrahedral complexes. In addition, the utility of the resultant complex was demonstrated in enantioselective

K. Endo, *Kinetically Controlled Stepwise Syntheses of a Heterometallic Complex and a Tetrahedral Chiral-at-Metal Complex*, Springer Theses,
https://doi.org/10.1007/978-981-16-1163-6_4

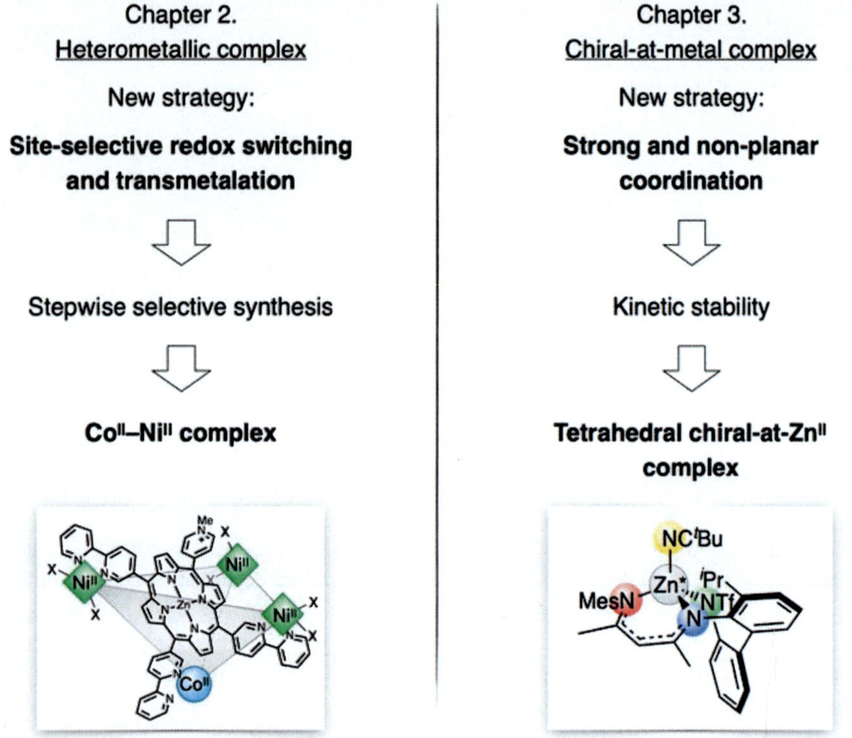

Fig. 4.1 Summary of Chaps. 2 and 3

catalysis. These results revealed the overlooked utility of the tetrahedral chiral-at-metal complexes composed of achiral ligands.

The expanded scope of kinetically controlled stepwise synthesis will be useful in preparation of metal complexes with new structures, which can be suitable for a particular purpose such as catalysis or material applications.

Curriculum Vitae

Kenichi Endo

Postdoctoral researcher, Teranishi Group,
Institute for Chemical Research, Kyoto University
Gokasho, Uji, Kyoto 611-0011, Japan
endo.kenichi.5m@kyoto-u.ac.jp
https://www.scl.kyoto-u.ac.jp/~teranisi/index_E.html
ORCID: 0000-0002-4128-6514

Education

- April 2013—March 2015: B. Sc. in Chemistry, The University of Tokyo, Japan, supervisor: Prof. Mitsuhiko Shionoya.
- April 2015—March 2017: M.Sc. in Chemistry, The University of Tokyo, Japan, supervisor: Prof. Mitsuhiko Shionoya.

K. Endo, *Kinetically Controlled Stepwise Syntheses of a Heterometallic Complex and a Tetrahedral Chiral-at-Metal Complex*, Springer Theses,
https://doi.org/10.1007/978-981-16-1163-6

- October 2018—December 2018: research internship, Philipps-Universität Marburg, Germany, supervisor: Prof. Eric Meggers.
- April 2017—March 2020: Ph.D. in Chemistry, The University of Tokyo, Japan, supervisor: Prof. Mitsuhiko Shionoya.

Appointment

- April 2020: postdoctoral researcher, Kyoto University, Japan, PI: Prof. Toshiharu Teranishi.
- Research topic: Surface decoration of semiconductor nanoparticles using surface coordination of metal complexes.

Research Interest

Exploiting the catalytic power of metal elements by controlling the chemical structures from coordination sphere to bulk heterogeneity.

Major Awards

- School of Science Encouragement Award | Graduate School of Science, The University of Tokyo (03/23/2020).
- The Best Student Presentation Award | The 69th Conference of Japan Society of Coordination Chemistry (09/23/2019).
- Student Presentation Award | The 98th Chemical Society of Japan Annual Meeting (03/23/2018).

Fellowship

- April 2017—March 2020: Research Fellowship DC1 | Japan Society for the Promotion of Science.